Building our Digital World:

DELIVERING SOFTWARE AND SYSTEMS IN SUPPORT OF OUR DIGITAL ECONOMY

Building Our Digital World Tools and Templates

Improving quality and consistency of outcomes in software development and agile processes

BY CECIL GARY RUPP

To my wonderful wife, Carolyn, and to our family, who have all demonstrated remarkable patience and sacrifice throughout the lengthy process and countless hours required to bring the Building Our Digital World series to life. I love you all.

Cover Artwork

Along the way to publishing this series of books, I discovered two important things. One, it takes a good team to develop a quality product. Second, even seemingly trivial tasks, like designing the book covers, can turn into large projects in their own right. Such was the case with designing the Building our Digital World Series book covers, and the artwork titled *Engaged in Our Digital World.*

The design concept began with the notion that we are all connected to a digital world; a world that is built by the efforts of software and IT-based systems developers. My book designer, Adam Hay, of Adam Hay Studio, and I saw several artificial intelligence (AI) based artworks that suggested such connections. While I liked the imagery of some of the AI-based artworks, they typically included pictures of androids or robots. I wanted to make the point that it's humans who are increasingly connected to our digital world. Besides the digital connections, I also wanted to convey the role software programming plays in forming such digital connections.

We could not find an existing commercial illustration and graphic that evoked the image I was seeking. So, I turned to the folks at Unzipped Designs, Victoria Moore and James Neely, to see if they could come up with a custom design for us. Victoria and James both work as artists in 3d modeling formats. James is a 3D creature and character artist who holds two patents for digital sculpting in product designs and the use of 3D models for pre-visualization of product packaging concepts for marketing purposes. Victoria is a software programmer who also develops 3d avatars for gamers. A perfect fit!

The final artwork displays a couple walking together with a digital rain immersing them in the virtualized digital world. The two individuals are avatars, who represent real-world humans in their digital interactions within this 'other' world. The programming code, lying upon the water-soaked ground, provides the foundation for their digital interactions.

Are the human-directed avatars and their digital world real, or imagined? I submit it's increasingly difficult to tell. After all, who as a gamer doesn't believe that the avatars they use, and those that surround them in their digital worlds, at least for the time of their play, represent reality and a real place where they reside and interact with others? I hope you enjoy the artwork as much as we enjoyed collaborating on the design.

Draft Copy Editor: Linda Orlando
Book Design: Adam Hay of Adam Hay Studio
Cover Artwork: Victoria Moore and James Neely of Unzipped Designs
Proofreader: Joe McDaniel of Book Crafters
Final Production Editor: Alexandra O'Connell of Your Resident Wordsmith
Marketing Specialist: Mary Walewski of Buy The Book Marketing
Website Design: Victoria Moore of Unzipped Designs

ISBN: 978-1-7341478-4-1

First Edition: January 2020

1 2 3 4 5 6 7 8 9 10

Contents

Author's Introduction 13

WHAT ARE SDLC / AGILE TOOLS AND TEMPLATES? 15

AUDIENCE 16

SCOPE 16

OBJECTIVES 16

Module 01. Agile Tools and Templates 17

Characteristics of Agility 19

Organizing Work Under the Agile SDLC Model 20

ENVISION PHASE 23

Overview 23

Forms & Templates 24

Project Charter 25

Vision Statement Template 26

Risk Management Register 28

Project Data Sheet 30

SPECULATE PHASE 33

Overview 31

Forms & Templates 34

Theme / Epic Structure Template (worksheet version) 34

Theme / Epic Structure Template (graphics version version) 36

Planning Poker Companion Template 38

Story Points Allocation Template 40

Release Plan 42

Sprint Planning & Meeting Schedule 44

Sprint Backlog Status (worksheet version) 46

Sprint Backlog Status (graphics application version) 48

14-Day Sprint Backlog - Capacity Template 50

30-Day Sprint Backlog - Capacity Template 52

Product Backlog Grooming Template 54

Iteration Retrospective Template 56

EXPLORE PHASE **57**

Overview 57

Forms & Templates 58

Story Completion Checklist 58

ADAPT PHASE **61**

Overview 59

Forms & Templates 62

Sprint Burnup Worksheet and Chart 62

Sprint Burndown Worksheet and Chart 64

Sprint Velocity Chart 66

CLOSE PHASE **68**

Overview 68

Module 02. Traditional Tools and Templates **69**

Author's Introduction 69

Characteristics of the Traditional SDLC Model 71

Organizing Work Under the Traditional SDLC Model 73

Contents

PROJECT CONCEPT PHASE **76**

Overview 76

Forms & Templates 78

Project Charter Template *78*

Business Case Document *80*

Feasibility Study *82*

Value Proposition Template *84*

Project Cost Benefit Analysis Template *86*

Concept of Operations (CONOPS) Template *88*

PROJECT START PHASE **91**

Overview 91

Project Initiation Agenda *92*

Project or Issue Submission Form *94*

Project Team Definition *96*

Stakeholder Identification List *98*

On-Boarding Checklist *100*

Initiate Project Checklist *102*

Project Resource Plan (Resources by Resource Type) *104*

PROJECT PLANNING PHASE **107**

Overview 108

Forms & Templates 108

Program Management Office (PMO) Checklist *108*

Project Management Plan Template *110*

Statement of Work (SOW) *112*

Project Approval Document *114*

Cost Estimating Worksheet *116*

Development Estimating Worksheet *118*

Project Capital vs. Expense Costs Worksheet 120

Configuration Management Plan 122

Risk Information Data Collection Form 124

Risk Analysis Template 126

Procurement Plan 128

Quality Plan 130

Project Organization Chart 132

Roles and Responsibilities Matrix 134

Required Approvals Matrix 136

Activity Worksheet in Work Breakdown Structure (WBS) Dictionary Form 138

Work Breakdown Structure (WBS) Resource Planning Template 140

Work Breakdown Structure (WBS) Template 142

COBIT Checklist and Review Template 144

COBIT Objectives and Audit Activity Report 146

Request for Information (RFI) Template 148

Root Cause Analysis Template 150

Project Plan Schedule Template 152

Project Manager's Weekly Checklist 154

List of Opportunities Summary Template 156

REQUIREMENTS PHASE **158**

Overview 158

Forms & Templates 160

Managing Scope and Requirements Checklist 160

Simple Approach to Requirements and Systems Analysis Template 162

Business Requirements Document (BRD) Template 164

Business Requirements Presentation to Stakeholders Template 166

Functional Requirements Document (FRD) 168

Contents

Use Case Template 170

Requirements Inspection Checklist 172

Requirements Traceability Matrix (RTM) Template 174

Requirements Changes Impact Analysis Form 176

Training Plan Template 178

Training Request Template 180

Service Level Agreement Template (SLA) 182

SYSTEM DESIGN PHASE **184**

Overview 184

Forms & Templates 186

Software Architecture Plan (SAP) Template 186

Systems Requirements Specifications (SRS) Template 188

Analysis and Design Document 190

Application Development Project Checklist 192

Technical Requirements Document (TRD) 194

Database Design Document 196

Website Planning Checklist 198

User Interface (UI) Design Template 200

Report Design Template 202

Code Review Checklist 204

Conversion Plan 206

TESTING PHASE **208**

Overview 208

Forms & Templates 210

Documentation QA Checklist Template 210

Test Scenarios Template (a.k.a. Building Test Scenarios) 212

Test Plan Template 214

System Quality Assurance (SQA) Checklist 216
Website Testing Summary Template 218
System Test Plan 220
User Acceptance Test Plan (UAT) Template 222
Testing Bug List 224
Testing Bug Report 226
Regression Testing Plan Template 228

PROJECT ADMINISTRATION PHASE **231**

Overview 231
Forms & Templates 232
Change Management Log 232
Action Item (Status) Log 234
Risk Management Log 235
Issue Identification and Resolution form. 238
Issues Management Log 240
Project Milestone Status Form 242
Project Status Report 244
Meeting Summary Template 246

DEPLOYMENT PHASE **249**

Overview 249
Forms & Templates 252
Process Guide Template 252
Installation Planning Guide 254
Software User Guide 256
System Administration Guide 258
Operations Guide 260
Production Implementation Plan (PIP) 262

Production Turnover Approval Form 264

PROJECT CLOSURE / MAINTENANCE PHASE **267**

Overview 267

Forms & Templates 268

Lessons Learned Template 268

Transition Out Plan 270

Post Project Review Survey Questionnaire 272

Installation Back-Out Rollback Plan Template 274

Project Acceptance Document 276

Modification / Change Control Request Form 278

Disaster Recovery Plan Information Template 280

Certificate of Compliance and Acceptance of Deliverable Form 282

Request for New Application / Enhancement Form 284

Product Retirement Plan Template 286

Off-Boarding Checklist 288

Global Application Support Summary Template 290

Developer Knowledge Transfer Report 292

INDEX **294**

BIBIOGRAPHY **297**

ACKNOWLEDGMENTS **298**

Figures

Figure 1.	Basic Agile/ Iterative Development Workflow Diagram	18
Figure 2.	Traditional SLC Model (A.K.A. Waterfall)	72
Figure 3.	SDLC Taxonomy in Directory Folder Structure	73

Building Our Digital World
Tools and Templates

Welcome! This book is the third book in a series of books on software and systems development practices, methods, tools and templates. The series title is *Building Our Digital World: Delivering software and systems in support of our digital economy.* The four books in the series include:

SDLC Foundations
An introductory and historical guide to software and systems development

Frameworks for Agility
A practical guide to achieving IT process maturity and agility in a constantly evolving industry

Tools and Templates
Improving quality and consistency of outcomes in software development and agile processes

DevOps
Communicating, collaborating, leaning, integrating and automating across development and operations processes

Through and through, I'm a process and tools guy. More precisely, I'm a methods and tools kinda guy. I can't help it. My earliest experiences in software development involved working in companies that applied a lot of rigor to their development practices, and the engineers I worked with were always kind enough to explain the methods and tools they preferred to use, and why. If brains can be wired to think in a specific way - based on early experiences, then my mind was wired to think in terms of *processes*, *methods*, and *tools* to support the performance of those types of activities.

A process is simply defined as a series of activities and tasks performed in a specific order to accomplish some goal. Under this definition, a recipe within a cookbook is a type of process. However, in this book, a process is a description of some type of work performed in software and systems development and operational support. And, there are many potential processes necessary to support a product across its life cycle, from initial concept through its end-of-life.

Activities and tasks are lower-level descriptions of work within a process. Readers of my other Building our Digital World (BODW) books know that I define activities as a summary level record of work - mostly for cost, time, and resource reporting purposes. In contrast, tasks are the lowest level of work defined within a process with a quantifiable deliverable output or result.

There can be any number of ways to potentially perform the underlying activities and tasks. Each approach represents a specific method. Not all methods are as efficient or widely used as others, and some methods are appropriate in some situations but not others. Tools are simply a device to simplify, make more efficient, and improve the quality of performing one or more methods.

The software industry as a whole produces more than its fair share of methods and tools. Quite a few years ago, I was given the title of Director of Methods and Tools within a very large IT-based professional services organization. The job of my team was to define and deploy professional service offerings as industry-specific solutions (e.g., retail, government, and financial industries) that our consultants could offer to our clients around the world. The challenge for our organization was to ensure the quality and consistency of our company's IT-based professional service offerings no matter where our customers were located.

One of the things I learned from that experience is that consultants and software engineers who have developed skills and preferences around specific tools, or in favor of specific methods they know, were not always receptive to learning new methods and tools. I learned that having acceptable alternatives is not a bad thing so long as the final products meet our customers' needs. On the other hand, the organization must balance their investment costs in supporting multiple yet redundant methods and tools.

I mentioned two paragraphs above that the IT industry as a whole produces more than its fair share methods and tools. For that reason, it is impractical to have a book that provides detailed instructions on every method and related tool available to the software development community. Fortunately, that's not an issue, as the number of methods that have widescale acceptance is relatively small. Still, this book includes more than 100 defined methods and tools that have found some level of industry acceptance.

The approach I took to deciding which tools and templates to include was fairly simple. I reviewed commercially available software/ systems development life cycle (SDLC) tools and templates offerings, and those used by my customers, to see what set of methods and tools are most commonly in demand.

During my research, I approached the number of SDLC and Agile forms and templates providers, reviewed their offerings, and ultimately partnered with **Documentation Consultants**, the developer of SDLCforms (www.sdlcforms.com). They have been in business since 2007 and have built a comprehensive library of SDLC and Agile tools and templates most asked for by their commercial and federal agency clients.

Not included in their offerings were a handful of methods and tools that I've used through the years and found highly valuable. The principals at SDLCforms agreed to add these forms and templates to their library. For example, many of the project management and several agile templates were developed through this collaboration.

This book provides information on the application of each method, with descriptions of the attending forms, checklists, reports, and templates. Each description includes the names of fields and the type of information managed in each. There are also representative graphic images of the forms and templates so that readers can get a feel for what the finished product might look like.

This book is split into two parts, one for agile-based tools and templates, and the other for traditional tools and templates. While I am an advocate of agile-based development practices, the reality is many organizations are still using traditional development practices, plus many of the traditional tools and templates support good engineering practices.

This approach ensures that my readers, regardless of whether they are using traditional, good engineering, or agile practices, or more often a hybrid approach that encompasses bits of all three practices, will find something useful in this book. Likewise, no two projects are alike, and the tools and templates that are useful on one project may not be the right set of tools and templates to use on the next.

Moreover, you will find multiple tools that help accomplish similar tasks. For example, a business requirements document, a functional requirements document, a use case, and epics and stories all facilitate the capture and documentation of business process requirements. This book includes templates for all those requirements gathering techniques.

Even if you are a diehard Agilist using epics and user stories to capture requirements, you may eventually run into a project involving enhancements to legacy products where the original development teams documented requirements using these other techniques. Moreover, you may join product development teams that employ different strategies than your preferences.

For all my readers who are advocates of agile, I recognize there is an elephant in this room. After all, the *Agile Manifesto* notes that we should value: [1]

- **Individuals and interactions** over processes and tools
- **Working software** over comprehensive documentation
- **Customer collaboration** over contract negotiation
- **Responding to change** over following a plan

The elephant is the use of processes, methods, tools and templates while remaining true to the values and principles outlined within the Agile Manifesto.

First, let me explain that I have no issues with any of these values. They are the cornerstone to efficiently and rapidly building software products that our customers will value. They also represent an ideal situation, even as they do not provide a prescriptive description of how work is performed under agile-based projects.

The *Independent Signatories of The Manifesto for Agile Software Development* represented at least 10 distinct so-called "light-weight" software development methodologies.[2] While there are overlaps in many of the approaches, such as iterative development, prototyping, and frequent customer feedback loops, there were also sufficient differences that the signers of the Agile Manifesto did not attempt to specify a representative agile process or methodology.

If you wanted to implement a specific agile approach, you would need to know something about the methods and tools employed by the approach you chose. And, that leads us back to a discussion on the value of defining methods and tools as preferred practices for each approach. This book does not attempt to describe common SDLC and Agile approaches. Instead, the goal is to provide descriptions of methods and tools that support the widest number of SDLC and Agile approaches.

WHAT ARE SDLC / AGILE TOOLS AND TEMPLATES?

The tools and templates described within this book are forms based. As I noted previously, the purpose of the tools and templates, as described in this book, is to facilitate the execution of industry-accepted methods in support of software and systems development.

In contrast, there are other software-based tools used in software and systems development, and I want to make sure my readers understand the difference in fit and purpose. For example:

- Integrated development environments (IDE) - support the development of software code, usually around one or more programming languages.
- Database tools - support the management of data within and across applications and for business analysis and reporting purposes
- Middleware tools - simplify application and system integrations.
- Source Control Management (SCM) tools – aid in the versioning and configuration management of artifacts under development and released into production, such as requirements documentation, architecture and design documents, software code, and test scripts, test scenarios, and test results.
- Continuous Integration and Continuous Delivery tools –support the automation of testing and provisioning of applications, security, and networking and computing resources.
- Application Lifecycle Management (ALM) tools – facilitate communications, collaboration, and lifecycle management of artifacts for a product across its life cycle.

I discuss these other types of software-based development tools within my first two Building our Digital World books, *SDLC Foundations, and Frameworks for Agility*. The upcoming BODW series book on DevOps will go into much greater detail on the use of software development tools to support Lean-Agile concepts through integration and automation of software and systems life cycle development and operational support processes.

AUDIENCE

I wrote this book for program managers, technical writers, developers, and quality assurance personnel supporting an IT product across its life cycle. Executive sponsors and other IT project stakeholders may find value in this book when they need to understand how or why work is performed in a specific manner, or when they need to understand what type of information is required to support the work.

Note: This and the other BODW books in the series are geared toward assisting practitioners and executives in larger business and government agencies that have multiple and simultaneous development activities and complex software and systems development needs. Still, you may find value in reading these books if you are an IT specialist working in a small company environment as the means to expand your professional knowledge.

SCOPE

The scope of this book is to introduce a series of methods and related forms-based tools used to help ensure the consistency and quality of work across a software and systems product life cycle. This book is divided into two modules. The first module introduces agile-based methods and tools, while the second module has a focus on traditional methods and tools. It is not practical to describe every possible method and tool in use in the IT industry. Instead, this book includes descriptions and examples of some of the most commonly used SDLC and Agile methods and tools.

OBJECTIVES

This book provides a library of methods and tools commonly used in both agile and traditional software development practices. While the book can certainly be read from front to back as an introductory guide, the main purpose is to provide quick access to descriptions of useful methods and supporting tools, on-demand, when and as required.

It's important to remember that most projects only require a relatively small subset of the methods and tools described in this book. Project managers, scrum masters, product owners, and development team members can select the tools they need independently or as a team during each product development engagement.

Module 01. Agile Tools and Templates

AUTHOR'S INTRODUCTION

This first module within this book includes a section on agile-based methods and tools. Before we get into describing the Agile-based tools and templates, I want to address that elephant in the room that I mentioned in the book's Author's Introduction section. That elephant is the use of processes, methods, tools and templates while remaining true to the values and principles outlined within the Agile Manifesto. So, let's take a few moments to discuss each value to fully understand their implications in terms of the content presented in this book.

Individuals and interactions over processes and tools.
The objective here is to ensure development teams understand that working with people, and how they work with people, is more valuable than the specific processes and tools they might use. The individuals and interactions values help ensure the team stays focused on the customer and that they work well together in a collaborative environment. There is no room for ego or disrespect when dealing with team members and customers to achieve the goals of the project.

Jim Highsmith, in his historical description of the Agile Software Development Manifesto, notes that agile is about the "*mushy stuff*" of values and culture." [3] The values of agile promote "trust and respect" with organizational models based on people and collaboration and help create environments where people want to work.

This value does not mean processes and tools go away, just that they are less valuable than how people work together. If the focus is on processes and tools the dev team may lose sight of why they're doing the work in the first place.

This first stated value is not meant to imply that processes and tools don't add value; they do. In my experience, processes and tools help improve the quality and consistency of work. This is especially so when those processes and tools help implement proven methods to accomplish product development and operations-oriented work.

I have one last comment on this subject. Defined methods and tools are guides that provide starting points to implement successful practices. I have almost always modified the methods and related templates I've used situationally to address the needs of the current project.

No two projects are the same. Customer needs are different. Different types of products and priorities influence development strategies. And, organizations do not invest in the same technologies, and often dictate preferred development strategies. In the end, the development teams need to assess the scope of work in context with their customers' constraints and figure out how best to proceed. Modification of standard practices is the norm and not the exception.

Working software over comprehensive documentation.
No matter how great the quality of documentation produced by a team, it's all for not if the software doesn't work, is buggy, or does not support the customer's current needs. The focus always has to be on developing quality software the customers want and can afford.

But that does not mean to imply that documentation is all bad. Once the development team moves on the products must be supported, maintained, and enhanced. There have been several projects I've been asked to lead where admin, development, and support documentation was never produced or long lost. Such situations always lead to greater lifecycle sustainment costs and early replacement. So, while we need the software to work in the first place, we also need sufficient documentation to ensure the product is sustainable across its life cycle.

Here's another issue to consider. Product sustainability costs can go out of sight when adequate documentation is not created and maintained on requirements, test scenarios, code, test results, and support, administration, and maintenance documentation. As a result, your paying customers may require development of detailed requirements, architecture, design, and operations-oriented documentation. Such a situation does not obviate the value of having working software over comprehensive documentation. It's really a matter of prioritization, timing and focus.

Customer collaboration over contract negotiation.
Here again, this value exists to ensure the development team always remembers the goal is to create software that customers will value. Under the traditional model, characterized as plan driven and following linear sequential processes, new releases of software typically span the fiscal year of activity. The whole world can change in a business within a year's time. Iterative development and prototyping practices reduce the development cycle time allowing more frequent, and customer directed releases.

But no one should kid themselves that contracts aren't important. Without a contract, or charter, the work cannot start. Whether you are bound by contract or charter, those agreements specify the high-level deliverables expected by those who are paying the bills, and the amount of money, time, and resources they can justify to the effort.

Any development team that thinks project constraints (e.g., scope, time, budget, resources, quality) are not important will quickly find their projects shut down if the sponsors or customers begin to feel the team is unable to achieve the specified deliverables within the approved constraints.

Responding to change over following a plan.
This value addresses the primary issue with following the traditional linear sequential development model. As I noted before, the priorities and needs within a business can change demonstrably over the course of a year. The traditional model typically enforced, though it did not have to, a strict process with distinct steps to define the product, discover and record the business and user requirements, define the product's architectures and designs, and then develop, test, and finally deliver a product to the customer or customers. Under the traditional model, short of a formal change review process, there was little ability to deviate from the original plans.

When I use the word process in this book it should not be confused with the word plan. Nor should the word process be interpreted to imply the product's entire lifecycle set of work as one process. The basic software development activities can take up to a year or more under the traditional software development model. Yet those same activities occur within the space of a weeks under agile-based timeboxed practices, and within a matter of days or even minutes when using continuous integration, testing, and delivering capabilities under modern DevOps-based practices.

Nevertheless, the SDLC activities remain pretty much the same thing. First, we decide what we need to build. Next, we gather requirements. Then we develop an architecture to specify the products components and a design approach to build a product or features within the product. Then we develop and test the product. And, finally, when we determine the product has enough functionality for release, and sufficiently bug free, we release it into production to our customers.

Principles of Agile
Besides the four values, the agile manifesto lists 12 principles that the authors followed. I won't bother to reiterate them here, as you can easily find them at the agilemanifesto.org website.[4] In my opinion, the main value of the principles outlined in the agile manifesto is they collectively describe a boundary of what's in and what's out within a methodology that purports to follow agile-based software development practices.

CHARACTERISTICS OF AGILITY

The signatories of the *Manifesto for Agile Software Development* use the word agile as an adjective defining an abstracted approach to software development. In other words, the thing we do is software development, but in an agile way. The noun form of agile is *agility* which, by definition, means the "ability to move quickly and easily. Since the agile manifesto concentrates on values and principles, a logical question should follow: "*how do we quickly and easily build software compared to the traditional model?*"

Virtually all agile practices implement a form of iterative development. The difference between agile development iterations and the linear sequential processes used in the traditional development model is mostly one of scale. They both include pretty much the same types of work practices.

However, agile-based practices break the work up into multiple yet smaller full-cycle iterations, with each development cycle increasing the product's value incrementally. In fact, the incremental approach is sometimes referred to as *iterative and incremental development (IID)*, to capture these two concepts.

Incremental development provides a number of advantages. Under the traditional model, developers often attempted to define all the business, user, and design requirements up front, and then develop and test the code in one relatively lengthy increment. The traditional approach led to lengthy development cycles and limited the ability to make major changes to the original plans once the development activities began. Moreover, the traditional model led to the development of multiple features simultaneously within a single development cycle. The single iteration development strategy of the traditional model made the programmers' code extremely complex and difficult to debug, adding to more potential delays in releases and out of control costs.

There are other practices associated with agility. For example, the elimination of unnecessary work (e.g., processes, activities and tasks) helps improve throughput, reduce complexity, and speeds the development process. As another example, the implementation of just in time (JIT) and Kanban concepts reduces work in process (WIP) and locks workflow in step with demand, which together serve to reduce complexity and bottlenecks in the development process. Readers who want to know more about these subjects should read the Frameworks for Agility and DevOps books within my *Building our Digital World* Series.

ORGANIZING WORK UNDER THE AGILE SDLC MODEL

As with the traditional model, there is not a specific industry standard for how agile-based work is logically structured, or even how to name the structures. There is not an overarching body that can make such a call. There are, however, two general strategies for organizing agile-based work. One strategy is to define Agile work as phases within each development iteration. The second approach is to define agile phases across the lifecycle of a product.

Important to note that the term phase should not imply a specific sequence within the software development process. That notion is a holdover from the traditional model. Rather, the term phase is a description of work by type. And, by the way, this definition for the word *phase* is also true for the traditional model. Long gone are the days where software developers follow a strictly linear sequential process from beginning to end through a products lifecycle. Or, at least those days should be long gone.

Given the iterative nature of software development underpinning most agile-based methods, the organization and classification of work is often depicted in some sort of circular shape. An example of a typical Agile model is shown in Figure 1. *Basic Agile/ IID Process with initial Requirements & Design Phase Work.*

In the Agile model depicted in Figure 1, the initial process kicks off with a Concept Phase that can include a lot of work that comes directly from the concept phase within the traditional model. In fact, it's not unusual for organizations to employ hybrid SDLC models that incorporate processes from both agile and traditional SDLC models. Activities within the initial concept phase would be one example.

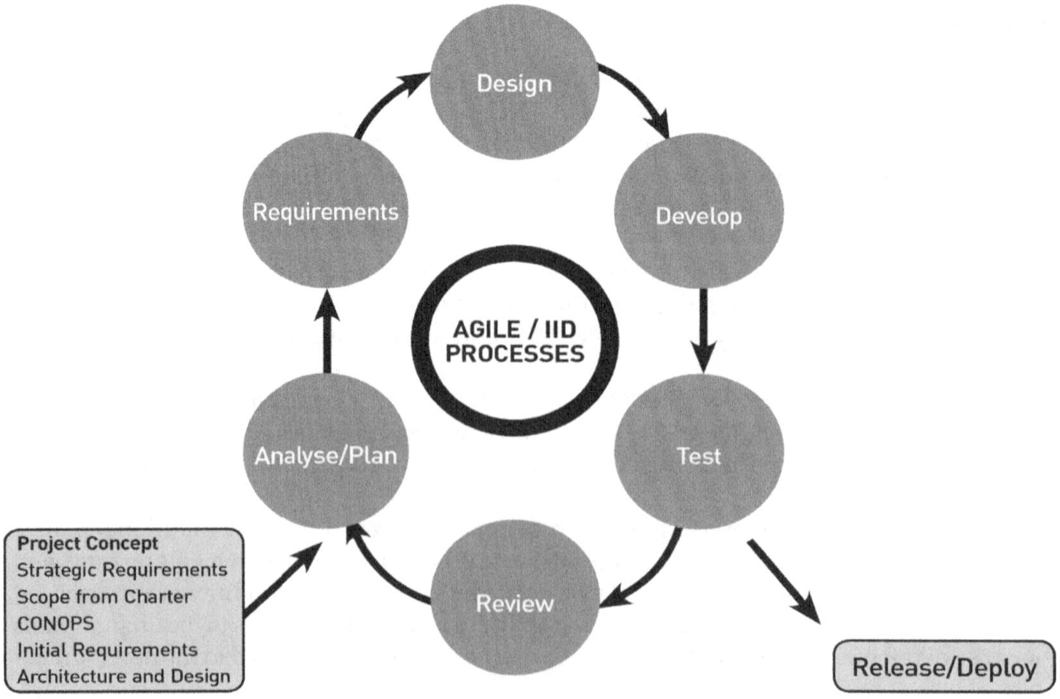

Figure 1. Basic Agile/ Iterative Development Workflow Diagram

There are work activities within the concept phase that are unique within the agile model. For example, there will be a product owner who will work directly with the customer and users to define their needs, usually written in the form of epics and stories. A business analyst and possibly members of the development team may assist in the gathering of these requirements. However, the format of documenting epics and stories is much more streamlined than the traditional model's business requirements and functional requirements documentation.

As the epics and stories are developed, the product owner will work with the development team to define priorities for each of the identified requirements. The document that lists those epics and stories in priority order is called the product backlog. In addition, the development team may work with the product owner to identify development strategies where from a development perspective it's logical to group specific requirements together. Those groupings are often referred to as themes.

Also, some teams may decide to do a high-level assessment of architectural and design requirements during the concept phase in an agile engagement, whereas those activities would fall into a separate and distinct phase within the traditional SDLC model.

Once the development team starts to work on constructing the product the type and flow of work is similar to the traditional model. The primary difference is the total cycle of product development work occurs more rapidly, repetitively, and more often under the iterative development practices of agile. This is why those depths are shown to occur in a circular fashion.

I noted at the beginning of this section that another strategy to organize agile-based work is to define agile phases across the lifecycle of a product. Again, there is no industry standard for how to do so. Within this book, and also true for my partners and SDLCforms, we have decided to organize agile-based work using the 5 Phases of An Agile Project Management (APM) Framework, as defined by Jim Highsmith in his book titled *Agile Project Management: Creating Innovative Products.*[5]

As a long time and certified project manager this type of work classification structure appeals to me. But there is no reason why an organization can't use some other identified agile standard or make up their own approach on how they want to organize and classify work within an agile framework.

The five APM phases include:

- **Envision**
- **Speculate**
- **Explore**
- **Adapt**
- **Close**

Each of the following chapters within this module map to one of the APM phases, and each chapter includes a description of the phase along with descriptions of the forms and templates associated with the APM phase.

If the organization engages in a quality improvement initiative, such as the Capability Maturity Model Integration (CMMI), they will have documented their processes (and underlying activities, tasks, methods, and tools) to help improve performance and consistency across the organization. However, as noted previously, documented methods and tools serve as a starting point. The same is true for the methods and tools identified in this book.

The practical usefulness and fit of the forms, templates, checklists or reports, described within this book, are highly dependent upon project complexity, management's work philosophy, auditing requirements, and government regulations. So, users should be prepared to tailor all

such artifacts to meet the needs of their organization and their projects.

Finally, not all methods and tools are appropriate to every project. It's much more likely you will only need a small subset of the methods and tools, described within this book, on any given project.

Envision Phase

OVERVIEW

Work within the Envision phase helps determine the product's vision and project scope, the project community, and how the team will work together. The term 'envision' is a clear departure from traditional phase names such as initiate and plan, which while subtle, is also significant. This is because, when envisioning a product, you accept that change is inevitable, and you are therefore ready to make any necessary adjustments. This is in stark contrast with the plan driven traditional model where there is the expectation the development team will follow prescriptive guidance as provided in the established plans.

The envision phase has the primary objective to develop the vision for the product. As a practical matter, this implies a discovery process that maps high-level business and user requirements to the corporate strategy, and as refined in the mission statement and portfolio objectives. The envision phase includes initial thoughts on what needs to be done (e.g., planned scope of work), who will do the work, and how the work is to be performed.

Without a product vision, the team lacks a roadmap that would help guide them to a successful outcome. Imagine taking a road trip from Los Angeles to Boston Massachusetts without a map. You might eventually get to Boston, but not without a lot of frustration, confusion, and misdirected side trips. You might also run out of money before you get to Boston. And, if you were taking the trip to attend an event in Boston, you might be so delayed that you miss it. This analogy works for software and systems development too.

Creating a useful vision takes work, discipline, and leadership. Hence, it is important there is an effective leader to define and articulate the vision. In an agile development project, that leader is typically the product owner. They represent the interests of paying customers, users, and stakeholders. Product owners refine the feature backlog in terms of their customer priorities. However, the development team is equally engaged in this assessment as component architectures, designs, and related development tasks impact implementation priorities.

I would also add that readers of this book look at the methods and tools described in the traditional SDLC model, and module to of this book, within the concept phase. The techniques developed in the traditional model still have relevance when defining user needs in support of the mission of the business, and the business case analysis that is necessary to justify the software and systems development and support investments.

During the Envision Phase, the product vision and priorities will change as new information is gathered. As a result, the vision should be reviewed periodically to ensure a clear understanding of direction and priorities by all those who are involved in the project.

All right; that's enough on describing the envision phase for now. Let's move on and get into the descriptions of the agile-based tools and template organized and managed within the envision phase.

Forms & Templates

PROJECT CHARTER

The project charter is not a novel concept. In fact, it's pretty much the same form that is used in the traditional model. The projects charter is essentially a contract between the executive sponsors or paying customers and the development team as to the work that must be performed.

As with any agreement, the project charter must describe the high-level deliverable items expected by the client, and the constraints the development team must live within. Those constraints include the authorized scope of work, the budgets allocated to the effort, the time frame in which the work must be completed, the resources that are assigned to the project, and descriptions and measurements defining what good looks like (e.g., quality objectives).

The Project Charter is developed early in the project lifecycle and clearly sets the stage by aligning the team and stakeholders by setting goals and expectations. In most cases, the Project Charter, in conjunction with the Vision Statement, is the vehicle provided to the project sponsors as a part of the approval process.

From a project management perspective, the charter supports the following activities:

- Identifies the initial vision, key product capabilities, and priorities for the project
- Defines key goals and requirements
- Captures and sets customer expectations
- Defines project participants and their roles
- Defines limits and constraints
- Establishes all resource needs and overall cost targets
- Creates a high-level view to the WBS and schedule
- Initiates negotiation and tradeoff
- Ultimately defines success

A general rule is to limit the project charter content to one page; however, as necessity arises, it may be necessary to provide the data on two pages. The Project Charter consists of the following data fields:

- Product Vision
- Mission Statement
- Success Criteria
- Target Objectives
- Key Deliverables
- Team Members
- Timeline
- Assumptions and Constraints
- Risks and Issues.

Project Charter
Project Name
Version

Note: In any table, select and delete any blue line text, then click Home→Styles and select "Table Text" to restore the cells to the default value.

Product Vision	Date	Rev.	Focus
Enter an abbreviated version of the product vision from the Vision Statement	*99/99/99*	*A*	*Enter focus of new project / product*

Mission Statement	Success Criteria
Enter a statement that describes the overall objectives of the project.	*Enter how you will measure success in achieving project vision.*

Target Objectives		Key Deliverables	
Finalize organizational structure	*99/99/99*	*Approved organization chart*	*Agile training complete*
Obtain budget approval	*99/99/99*	*Approved budget*	*Documentation deliverables defined*
Hire contract personnel	*99/99/99*	*Integration of team members*	
Final review of team members	*99/99/99*		
Train personnel in Agile	*99/99/99*		

Team Members		Timeline	
Name	**Responsibility**	**Actions / Milestones**	**Target Date**
Bill Gates	*Project Manager*	*Enter any actions or milestones, such as:*	*99/99/99*
Jessie Stewart	*Product Manager*	• *Initial meeting to develop high level requirements and team organization chart*	*99/99/99*
Joe Jones	*Marketing*	• *Organization chart and budget approval meeting*	*99/99/99*
Betty Barsam	*Sales*	• *Create user stories.*	*99/99/99*
Nancy Nansen	*Development*		

Assumptions and Constraints	Risks and Issues
Enter and assumptions or constraints that will impact the project	*Enter any risks or issues.*

Confidential – ©2016 Documentation Consultants (www.SDLCforms.com) Document: A2000 Page 1 of 1

www.SDLCforms.com

VISION STATEMENT TEMPLATE

A product vision provides a summary outline of the desired future state for a product when the product reaches maturity. The vision focuses on the business conditions addressed by the product and includes stakeholder views of necessary capabilities. The vision statement is specified in non-technical terms.

Note the distinction between capabilities expressed as stakeholder's key needs, as opposed to listing features and functions. Since the product is not yet developed, we can't really know how the desired capabilities will be delivered in the form of features and functions. So, the focus here has to be on defining what capabilities the product needs to deliver to users and customers.

The Vision Statement development effort is led by the product owner, with input coming from customers, potential end users, other important stakeholders, and of course the development teams. The development teams focus is on identifying capabilities that may not be intuitively obvious to customers and users and beginning to look into how to translate the desired capabilities into specific product features and functions.

The vision must be clear and easy to understand by all team members. The vision must support a common purpose and written with carefully selected verbiage so as to not be misunderstood by stakeholders and team members.

The Vision Statement consists of the following data fields:

- Title
- Target Group – Addressed by the vision statement
- Needs - How does the product create value for its customers and users?
- Product – A description of the product's use
- Business Goals – How will the product benefit the company?
- Competitors – List potential competitors and their strengths and weaknesses vis-à-vis the development of a new commercial or custom application for internal use.
- Revenue Sources - identifies the executive sponsors or other paying customers
- Cost Factors – drivers impacting development, marketing, support and maintenance costs
- Channels - if this is a commercial product, define how the product will be marketed, sold, and supported

Note that the vision statement form in the sample template form has only one page. In practical application, this template may extend across multiple pages.

Vision Statement
Project Name
Version

Note: In any table, select and delete any blue line text; then click Home→Styles and select "Table Text" to restore the cells to the default value.

Agile Vision Statement			
Title:	*Describe project / product purpose.*		
Target Group	**Needs**	**Product**	**Business Goals**
Which market segment does the project/ product address? *Who are the users and customers?*	*Define how does the project/product create value for the users?* *What problem(s) does the vision solve?* *What benefits does it provide?*	*Describe the product.* *What makes the product special and desirable?* *It is realistic and feasible to develop the project / product?*	*How will the project / product benefit the company?* *What are the specific business goals?* *What are the two or three most important goals?*
Competitors	**Revenue Sources**	**Cost Factors**	**Channels**
Who are the project's / product's main competitors? *What are the strengths and weaknesses of this product versus the competitor's product?*	*How can the project / product be monetized to generate revenue?* *What steps must occur to open up the revenue sources?*	*What are the main cost factors to develop, market, and maintain the project / product?* *Which resources and activities incur the highest costs?*	*What channels will you use to market and sell the product to the target group of customers?* *Identify the status of these channels. Do they exist or must they be created?*

www.SDLCforms.com

RISK MANAGEMENT REGISTER

Risk identification and mitigation strategies planning must start at the very beginning of the envision phase. Without this information it is impossible to have confidence that the project can be delivered within the constraints approved by the executive sponsors or paying customers.

Before explaining the purpose of this risk register, we need to make sure that we have a clear understanding of what a risk is and how risks differ from issues. By definition, risks are potential situations that can expose the organization to danger, harm, or loss. Risks can negatively impact the organization in many ways, such as operational, financial, security, performance, skills and resources. A proper risk assessment must address all of these and other potential problems that might arise when developing the product. As potential risks are identified, the team should identify potential mitigation strategies to minimize the impact. The typical risk mitigation strategies are avoidance, acceptance, reduction or control, and transference.

In contrast, an issue is a risk realized. Now, if the organization did not identify the risk in the first place that does not mean it wasn't first a risk. The risk was just not previously identified. When a risk becomes an issue, the organization must move to implement contingency plans. Where risk mitigation plans seek to avoid a risk from happening in the first place, contingency plans focus on what to do in the worst-case scenario.

The Risk Management Register provides a vehicle for recording potential risks on a project, by recording, as a minimum, a brief explanation of the risk, the proposed mitigating strategy, the probability of the risk, size of loss (in days), and risk exposure (in days), actions needed to resolve the risk, and the current status.

The Risk Management Register should be reviewed periodically by the project team to ensure everyone is constantly on alert to identify these and other potential risks. Under agile practices, the best time to do this is during the Sprint review meetings.

The following data fields are provided in the Risk Register:

- **Risk**
  - This brief explanation of the risk should be clearly understood by all team members.
- **Mitigation Strategy**
  - How does the team propose to mitigate the risk?
- **Date Identified**
  - The originating date.
- **Owner**
  - What department/person owns the risk?
  - Who is responsible to manage, control and take action in response to the risk?
- **Probability of Risk**
  - Generally, this is stated as a percentage from 1% to 100%.
- **Size of Loss**
  - The size of the loss is measured in days.
- **Risk Exposure**
  - The risk exposure is measured in days.
- **Action to Manage Risk**
  - What actions are underway to minimize or eliminate the risk?
- **Current Status**
  - Suggested statuses = "Open," "Closed," or "Under Review."

Risk Management Register
Project Name
Version

Note: In any table, select and delete any blue line text; then click Home→Styles and select "Table Text" to restore the cells to the default value.

Risk	Mitigation Strategy	Date Identifed	Owner	Probability of Risk	Size of Loss (Days)	Risk Exposure (Days)	Action to Manage Risk	Current Status
Utilizing new integration technology to connect to partnering systems.	• *Monitor frequently.* • *Search for potential new partners* • *Consider internal staff taking over function*	*07/23/16*	*IT*	*30%*	*15*	*12*	*Review alternative to new integration technology*	*Open*
Dependent systems are incompatible during integration	• *Consider other dependent systems* • *Consider middleware*	*08/14/16*	*IT*	*80%*	*12*	*9.5*		*Closed on 09/27/16*
There won't be time in the QA process to validate all possible combinations of cell phones	• *Develop plan to test only stories with high point values* • *Consider offering only most common phones*	*08/22/16*	*Marketing*	*87%*	*10*	*3*	*Consider using only 6 most popular phones.*	*Under Review*
Strong possibility that system performance will be severely degraded due to huge volume of transactions	• *Evaluate value of adding more servers.* • *Re-evaluate if transaction volume is factual*	*09/12/16*	*Operations*	*90%*	*20*	*12*	*Consider adding high-speed servers to enhance performance.*	*Closed*
Insufficient time will be allotted to provide end user documentation in 12 languages	• *Consider reducing number of languages*	*09/24/16*	*Technical Support*	*30%*	*2*	*2*	*Consider reduction from 12 to 6 languages.*	*Open*

www.SDLCforms.com

 Document: A4000 Page 1 of 1

PROJECT DATA SHEET

Created either in conjunction with development of the vision statement or shortly thereafter, the Project Data Sheet (PDS) offers a summary level view of information related to a product development project. The PDS quickly conveys to executive management, customers, project team members, and other critical stakeholders, important information about the project.

The PDS provides a concise overview of scope, schedule, resources key business and quality objectives, product capabilities, and project management information. Though it is a relatively simple document, its power comes from the concise format that is appealing to busy members of the project's stakeholders. During executive management, client, user, and product backlog reviews, the PDS provides a quick reminder of the strategic drivers that justified the investments in the project, and the resources and other constraints applied to the project.

The following data fields are provided in the Product Data Sheet:

- **Project Manager**
  - Identify the Project Manager.
- **Product Manager**
  - This is the product owner.
- **Executive Sponsor**
  - Identifies the
- **Project Start Date**
  - Identifies when the project started. E.g., the kickoff date
- **Clients**
  - List
- **Client Benefits**
  - Identify the major benefits the client has identified
- **Project Objective Statement**
  - Provide a short and concise
- **Performance Attributes**
  - Identify performance and quality aspects of the product.
- **Trade-Off Matrix**
  - For Scope, Schedule, Resources and Stability, identify whether the trade-off value of each is Fixed, Flexible or
- **Product Architecture**
  - Provide a high-level overview of the proposed product architecture (e.g., Windows 10, SQL Server. HTML, CSS, .NET, etc.)
- **Business Subject / Activity**
  - Identify the key business groups and their objectives.
- **Issues and Risks**
  - Identify
- **Major Project Milestones**
  - Identify critical deadlines for the project from inception to completion.

Project Data Sheet
Project Name
Version

Note: In any table, select and delete any blue line text; then click Home→Styles and select "Table Text" to restore the cells to the default value.

Project Manager	*Joe Manager*	**Executive Sponsor**	*Felicia Friendly*
Product Manager	*Betty Brian*	**Project Start Date**	*09/12/16*

Clients	Client Benefits
Risk Management	*Reduce manual data entry*
Commercial Operations	*Improve customer service*
Marketing	*Faster order processing*
Sales	*Better account management*

Project Objective Statement	Performance Attributes
1. The objective is to build a customer-facing website that	*100,000 transactions per day.*
to permit customers to order online.	*Available in 6 languages.*
2. The website will include customer history and price	*Training available in 6 languages not to exceed ½ day.*
reductions for volume sales.	
3. The system must be operational by no later than	
September 2017.	

Trade-Off Matrix

	Fixed	Flexible	Accept	Target
Scope		*X*		*25,000 FP*
Schedule	*X*			*+/- 6 weeks*
Resource		*X*		*+ / - $5 Million*
Stability			*X*	
Project Delay Cost Per Month	*$62,500*			
Exploration Factor	*7*			

Product Architecture
Windows 10
SQL Server
HTML5 , CSS and .NET
Maximum use of API's
Interface with sales & marketing system

Business Subject / Activity
Sales Management
Sales Analysis
Sales follow-up
Regional distribution of profits
Marketing
Lead generation
Lead follow-up
Order management
See the Feature Breakdown Structure for details.

Issues & Risks
Sales staff may be reluctant to embrace new system
Technology may not be available to support project
Team members may not agree on requirements baseline.

Major Project Milestones	
Develop website	*02/10/17*
Incorporate eCommerce capabilities	*04/23/17*
Provide profit distribution	*06/12/17*
Provide marketing features	*08/28/17*

www.SDLCforms.com

Speculate Phase
Overview

This phase is an accommodation to the understanding that no matter how rigorous our planning might be there is going to be unanticipated changes to customer needs, priorities, risks and issues that cannot be discovered until after the project starts. The initial plan is only as good as the information available at the time it is developed. When things change, we need to re-plan.

In the APM framework, the speculate phase is the equivalent of project planning in the traditional model. The process of speculating establishes a target and direction for the project team. However, the team also acknowledges that they expect to have to accommodate change over the lifetime of their project. The agile approach to planning deviates from the traditional model in that accommodation.

In other words, under the traditional software development model, there is little expectation that the planned and scheduled activities will change very much short of a major risk or issue that would justify a formal analysis and change request. Deviations within the traditional project and schedule plans are therefore viewed as negative in the traditional model, while changes to project plans are viewed as positive in the agile model.

The reason change is viewed positively in the agile model is that the shorter development iterations provide an opportunity to frequently test the pulse of the customer to see if their business needs or priorities have changed and redirect their work accordingly. If so, the development team updates the product backlog and planned sprint backlogs to support the new change requests and changes in priorities.

The outputs of the speculating phase include requirements, usually defined in the form of epics and user stories, and updates to the product backlogs and release plans. Instead of attempting to do all this requirements analysis and backlog planning work in one shot at the beginning the project, the speculate phase spans the entire set of development and release iterations. Updates include information on product specifications, platform architecture, necessary resources, risk analysis, defect levels, business constraints, desired capabilities and features and functions, development priorities, and target feature release schedules.

Product development features implement desired business and user capabilities. In other words, the development need must understand the needs of their customers, expressed as capabilities the product must provide, before they can design the architectural components, and feature and function designs to implement those capabilities.

More complex software systems include security, middleware, embedded software (e.g., firmware), hardware and networking capabilities. Implementation of these infrastructure-oriented components become part of the identified requirements and product backlog. The development team must work with the product owner to identify and prioritize these needs during the product backlog refinement meetings.

Forms & Templates

THEME / EPIC STRUCTURE TEMPLATE (WORKSHEET VERSION)

The basic structure to capture and organize requirements under Agile development practices includes Themes, Epics, Stories, Acceptance Criteria, and Features, defined as follows:

- Theme - is a collection of stories that somehow usually go together. For example, the collection of stories may describe an end-to-end business process. Or, the collection may be useful from an implementation perspective, such as including all stories that affect the design of a web page.
- Epic – Is a high-level story that includes multiple requirements. Those requirements must be broken down into more succinct user stories before development can start.
- Stories (a.k.a. User Stories) – should express exactly one technical, business or user capability requirement. The format is the most basic and succinct approach to writing a requirement, written as follows:
  - "As a <type of user> I want to <describe capability> so that <desired objective>"
- Acceptance Criteria – as an extension to user stories, specifies the conditions that will satisfy the users when the capability is implemented in the software application of system. The general format of an acceptance criteria is:
  - "Given some <condition>, when I do <some action>, then I expect <a result>.
- Feature – defined by the developers, features are the product solutions that implement the capabilities described in the themes, epics and user stories.

There is a hierarchical organization to how requirements are structured under agile practices. At the highest level are the themes which decompose into epics. Epics further decompose into stories. And, user stories decompose into the acceptance criteria that refines the understanding of what good looks like when the story is implemented within a feature.

The Theme / Epic Structure Template graphically depicts the hierarchical organization of themes, epics and user stories. The collections of stories under themes and epics helps organize and communicate work that logically goes together. In the product backlog, it's not unusual to see themes and epics listed before they are decomposed into user stories. But themes and epics must be decomposed into stories to specify the work required within the sprint backlogs.

The Theme / Epic Structure from SDLCforms is implemented in both Excel and Visio. While Excel is typically more available to users, the Visio version provides better "drag and drop" capabilities to move/copy objects within the document.

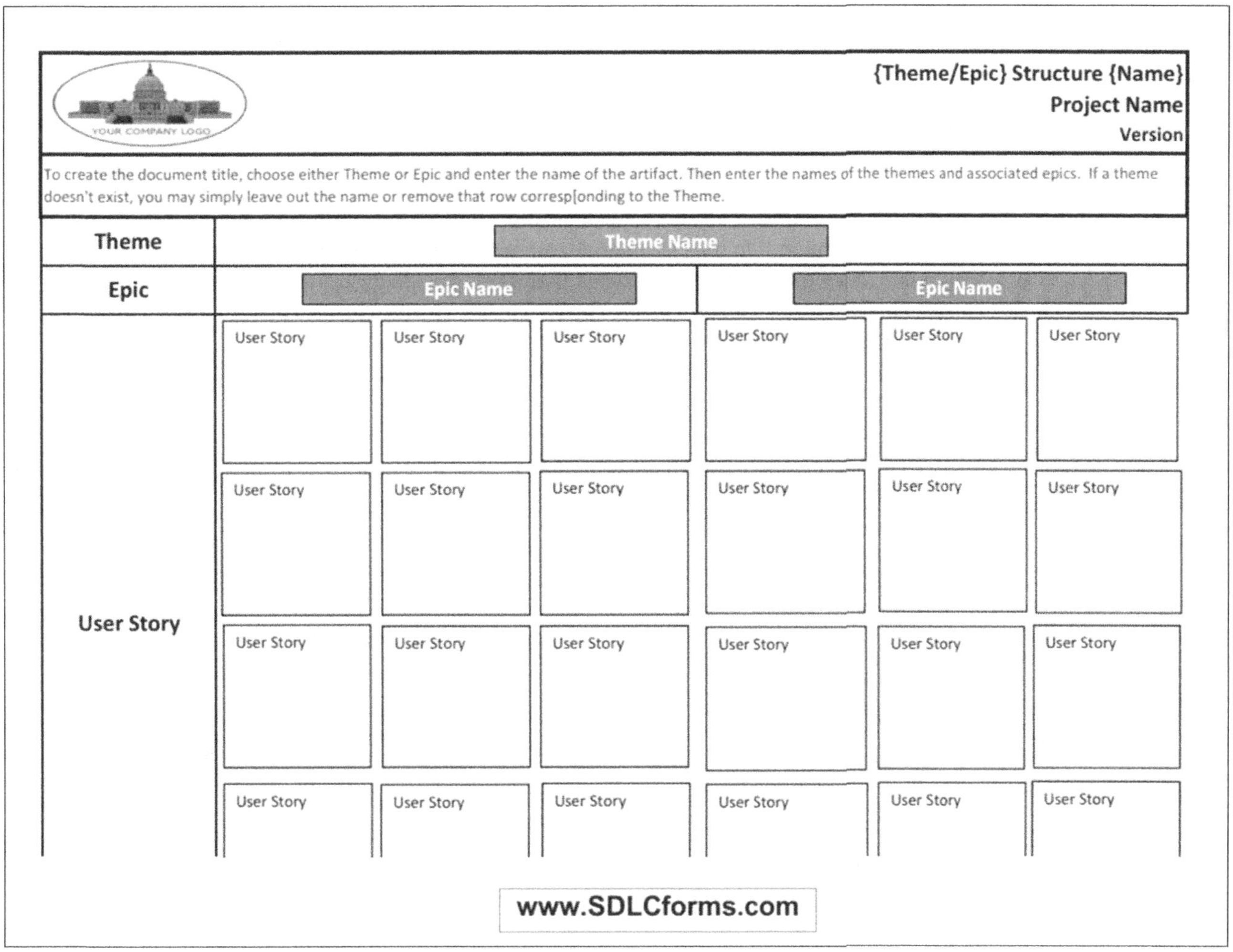

YOUR COMPANY LOGO

{Theme/Epic} Structure {Name}
Project Name
Version

To create the document title, choose either Theme or Epic and enter the name of the artifact. Then enter the names of the themes and associated epics. If a theme doesn't exist, you may simply leave out the name or remove that row corresp[onding to the Theme.

Theme	Theme Name					
Epic	Epic Name			Epic Name		
User Story	User Story	User Story	User Story	User Story	User Story	User Story
	User Story	User Story	User Story	User Story	User Story	User Story
	User Story	User Story	User Story	User Story	User Story	User Story
	User Story	User Story	User Story	User Story	User Story	User Story

www.SDLCforms.com

THEME / EPIC STRUCTURE TEMPLATE (GRAPHICS VERSION VERSION)

The basic structure to capture and organize requirements under Agile development practices includes **Themes, Epics, Stories, Acceptance Criteria**, and **Features**, defined as follows:

- Theme - is a collection of stories that somehow usually go together. For example, the collection of stories may describe an end-to-end business process. Or, the collection may be useful from an implementation perspective, such as including all stories that affect the design of a web page.
- Epic – Is a high-level story that includes multiple requirements. Those requirements must be broken down into more succinct user stories before development can start.
- Stories (a.k.a. User Stories) – should express exactly one technical, business or user capability requirement. The format is the most basic and succinct approach to writing a requirement, written as follows:
- "As a <*type of user*> I want to <*describe capability*> so that <*desired objective*>"
- Acceptance Criteria – as an extension to user stories, specifies the conditions that will satisfy the users when the capability is implemented in the software application of system. The general format of an acceptance criteria is:
- "Given some <*condition*>, when I do <*some action*>, then I expect <*a result*>.
- Feature – defined by the developers, features are the product solutions that implement the capabilities described in the themes, epics and user stories.

There is a hierarchical organization to how requirements are structured under agile practices. At the highest level are the themes which decompose into epics. Epics further decompose into stories. And, user stories decompose into the acceptance criteria that refines the understanding of what good looks like when the story is implemented within a feature.

The Theme / Epic Structure Template graphically depicts the hierarchical organization of themes, epics and user stories. The collections of stories under themes and epics helps organize and communicate work that logically goes together. In the product backlog, it's not unusual to see themes and epics listed before they are decomposed into user stories. But themes and epics must be decomposed into stories to specify the work required within the sprint backlogs.

The Theme / Epic Structure from SDLCforms is implemented in both Excel and Visio. While Excel is typically more available to users, the Visio version provides better "drag and drop" capabilities to move/copy objects within the document.

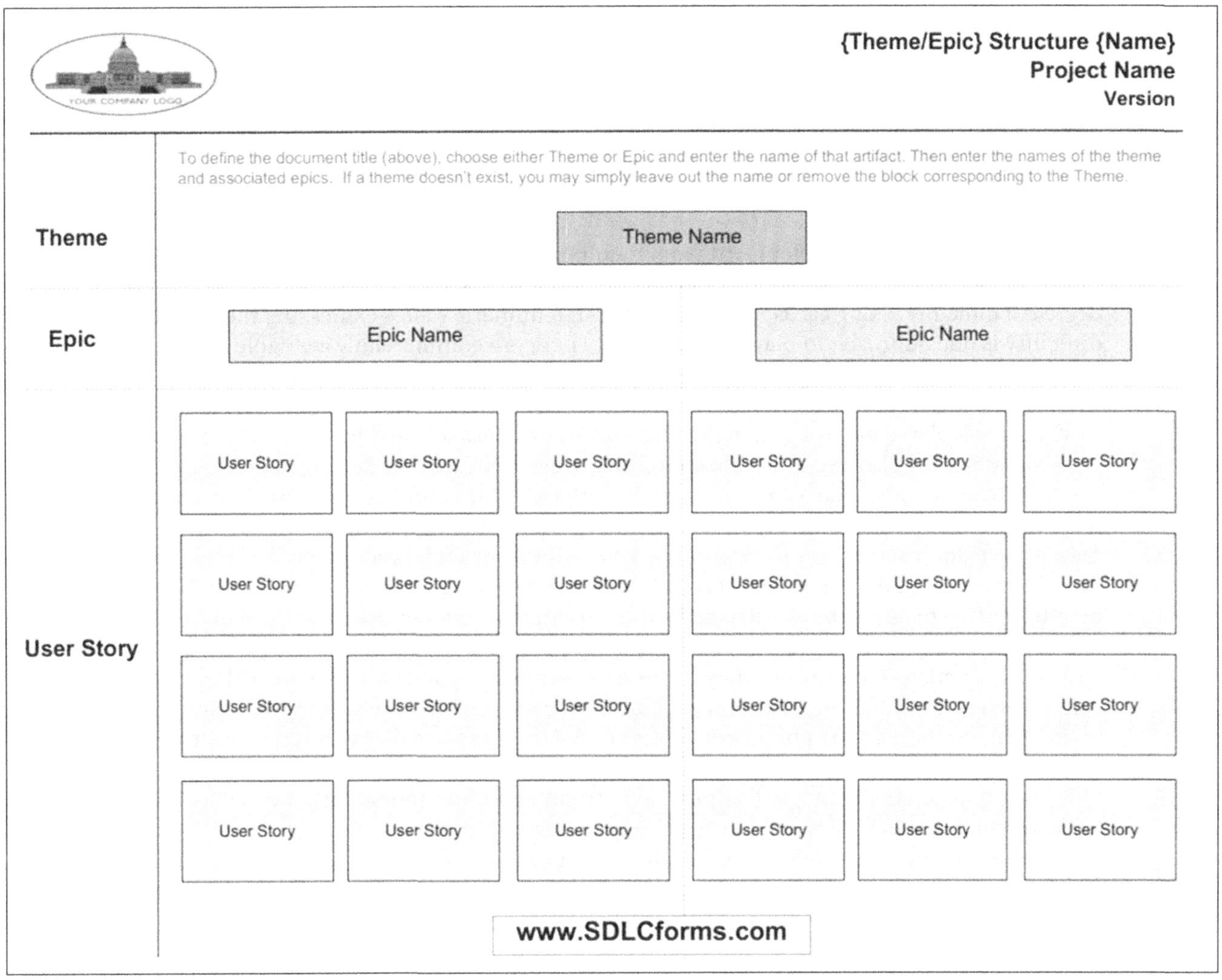
{Theme/Epic} Structure {Name}
Project Name
Version

To define the document title (above), choose either Theme or Epic and enter the name of that artifact. Then enter the names of the theme and associated epics. If a theme doesn't exist, you may simply leave out the name or remove the block corresponding to the Theme.

Theme

Theme Name

Epic

Epic Name

Epic Name

User Story

User Story	User Story	User Story	User Story	User Story	User Story
User Story	User Story	User Story	User Story	User Story	User Story
User Story	User Story	User Story	User Story	User Story	User Story
User Story	User Story	User Story	User Story	User Story	User Story

www.SDLCforms.com

PLANNING POKER COMPANION TEMPLATE

Estimating task durations turns out to be a fairly difficult task, even though that approach was common under the traditional model. Part of the reason estimate task durations is so difficult is because it's highly dependent upon the methods and tools used each time the task is performed, and the skills of the team members assigned to perform the task. Planning poker is an estimation technique that avoids these issues in a consensus building process that employs gaming techniques.

In this approach, the development team members meet to discuss upcoming tasks within the sprint and assess the work effort for each, not in terms of time estimates, but in terms of degree of difficulty. Using cards in the deck to assign numeric values expressing the degree of difficulty is one approach to playing this game. There are commercially available cards that employ numbers from the Fibonacci series, as another alternative. But some team members like to use things like T-shirt sizes; for example: small, medium, large, and extra-large.

The game is played by having team members initially discuss the work involved with each task. Team members are also free to ask questions to further clarify their understanding of the work. Next, the team members rank each of the tasks with the values they believe most appropriately represents the degree of difficulty for each. If there is a large discrepancy in the values that are played the team will further discuss the work to refine their understanding of the effort. The cards are played again until consensus is reached. If consensus cannot be reached, one or more members of the team will be asked to do further research to better understand the requirement and to determine the most practical development approach.

As the team plans their sprints, they assess how many story points worth of work they can do within each sprint duration. Based on backlog priorities, stories are assigned to the sprint up to the total number of story points worth of work the team expects they can accomplish. Upon completion of each sprint, the team analyzes how much work they accomplished in terms of total story points. Finally, they will adjust their estimates for how much work they can perform based on the trend in their weekly story point averages.

The *Planning Poker Companion Template* provides a tool to facilitate the planning poker method for estimating story points. The tool provides four optional methods to capture and display estimated values. E.g., as a Fibonacci Series, Numeric Series, Hours Approximation, and Days Approximation.

Averages are then calculated for:

1. Last Round Only
2. Average All Rounds,
3. Average Without First and Last Rounds.

Planning Poker Companion - {User Story}
Project Name
Version

User Story ID		Description	

Fibanocci Series (Rounds Column Values of 0,1,2,3,5,8,13,21,34,55,89)

	Rounds						Averages (See Instruction Sheet)			
Agile Team	1	2	3	4	5	6	Last Round Only	Average All Rounds	Average w/o First & Last	Priority
Product Owner							0.00	0.00	0.00	
Scrum Master							0.00	0.00	0.00	
Project Manager							0.00	0.00	0.00	
Agile Team # 1							0.00	0.00	0.00	
Agile Team # 2							0.00	0.00	0.00	
Agile Team # 3							0.00	0.00	0.00	
Agile Team # 4							0.00	0.00	0.00	
Agile Team # 5							0.00	0.00	0.00	
Agile Team # 6							0.00	0.00	0.00	

Numeric Series (Rounds Column Values of 1 through 10)

	Rounds						Averages (See Instruction Sheet)			
Agile Team	1	2	3	4	5	6	Last Round Only	Average All Rounds	Average w/o First & Last	Priority
Product Owner							0.00	0.00	0.00	
Scrum Master							0.00	0.00	0.00	
Project Manager							0.00	0.00	0.00	
Agile Team # 1							0.00	0.00	0.00	
Agile Team # 2							0.00	0.00	0.00	
Agile Team # 3							0.00	0.00	0.00	
Agile Team # 4							0.00	0.00	0.00	
Agile Team # 5							0.00	0.00	0.00	
Agile Team # 6							0.00	0.00	0.00	

www.SDLCforms.com

STORY POINTS ALLOCATION TEMPLATE

The Story Points Allocation Template provides a tool for recording story points that have been allocated to each user story, or more precisely to the tasks performed to implement the user story.

A user story articulates the requirements for a specific business or user need that is assigned to the software development team from the product and Sprint backlogs. As part of the sprint planning process, the development team meets to analyze the work required to implement the requirements defined in the user stories as features within the product. This is the point at which the development team defines the software development tasks that are required to implement those features.

Using estimations of the degree of difficulty as story points, as opposed to time duration estimates, has the advantage of assessing the amount of work that can be performed in terms of the overall capacity of the team. It may be difficult, for example, to estimate how long a particular feature will take to develop but relatively easy to understand if it is more complex than others, in which case it should be assigned more story points. In addition, less refinement is necessary to precisely define task durations when the task is assigned to team members with different levels of skill.

Elements considered in assigning a story point include the complexity of the story, the number of unknown factors and the potential effort required to implement it. Story points are usually expressed according to a numerical range, such as an adaptation of a Fibonacci sequence, the numbers in a deck of cards, t-shirt sizes. Though not typical in an agile planning process, the team members can assign estimated hours or days for each assigned task within the Planning Poker Companion Template.

Note that the template also allows the team members to designate the theme or epics that the user stories are assigned to. This is useful when the team is trying to organize work to expedite the development of features around the larger business and user requirements defined in the themes and epics.

Story Points Allocation
Project Name
Version

Note: In any table, select and delete any blue line text; then click Home→Styles and select "Table Text" to restore the cells to the default value.

Sprint	Epic or Theme (if applicable)	User Story	Points
1	*Develop online customer feedback*	*Customers can enter their name, email address and phone number to identify who they are.*	*6*
		Customers will be able to rate the products on a scale of 1 to 10.	*3*
		Customers will be able to enter comments on their ratings.	*1*
		Average ratings per product will be provided for Marketing.	*12*
		Marketing will have capability to respond to customer feedback	*3*
2	*Provide reports on customer feedback*	*Reports will be provided to Sales & Marketing averaging all product ratings.*	*12*
		Reports will be formatted in both printed form and as Excel spreadsheets.	*3*
		Reports will be provided by as weekly summaries and available on-demand.	*6*

www.SDLCforms.com

RELEASE PLAN

When developing software under an agile framework, the typical strategy is to ensure that the final deliverables are potentially shippable at the end of each Sprint. However, for logistical reasons, it's not always practical to release products frequently to the operations department due to training, support, and resource availability issues. In such cases the sprint deliverables accumulate for some period of time and then are released all at once in a formal release.

The product owner working with the development team will determine the release strategy for the product. For example, they may decide to release new products on a predictable timeframe, such as every three months. Or, they may decide it's more practical to release products with the implementation of a combined set of features and functions that support deployment of a new or enhanced business process.

A Release Plan communicates the release date, a description of the features scheduled for inclusion in the release, and a listing of the user stories implemented in the release.

The Release Plan is a collaborative effort involving the product owner, development team members, and scrum master. Release planning is technically part of the product backlog refinement process. As such, the release planning activity must evaluate technical feasibility and dependencies in terms of resources available and work capacity within the release cycle.

The Release Plan includes the following data fields:

- Release "Iteration" Number
- Sprints Numbers
- Sprint Start and end dates
- Collective story points for each sprint
- List of stories included in each iteration/sprint; color coded to indicate current status:
  - To Do items
  - In Process items
  - Done items
  - Backlog items.

YOUR COMPANY LOGO

Release Plan
Project Name
Version

Task / Story Legend: To Do | In Process | Done | Backlog

Iteration	Sprint	Start Date	End Date	Story Points	User Stories		
0	1	99/99/99	99/99/99	99	Task # Task Description	Task # Task Description	Task # Task Description
1	2	99/99/99	99/99/99	99	Story # Story Description	Story # Story Description	Story # Story Description
1	3	99/99/99	99/99/99	99	Story # Story Description	Story # Story Description	Story # Story Description
1	4	99/99/99	99/99/99	99	Story # Story Description	Story # Story Description	Story # Story Description
1	5	99/99/99	99/99/99	99	Story # Story Description	Story # Story Description	Story # Story Description
2	6	99/99/99	99/99/99	99	Story # Story Description	Story # Story Description	Story # Story Description
3	7	99/99/99	99/99/99	99	Story # Story Description	Story # Story Description	Story # Story Description

www.SDLCforms.com

SPRINT PLANNING & MEETING SCHEDULE

The *Sprint Planning & Meeting Schedule* provides a tool to record the status of sprints and associated tasks over a specific period of time.

Sprint planning is a collaborative meeting attended by Scrum Master - who facilitates the meeting, a Product Owner - who clarifies the details of the product backlog items, their respective acceptance criteria, and priorities, and development team - who define the work in the form of tasks, and the effort, in the form of story points, necessary to meet Planned commitments for each Sprint.

During each sprint planning meeting, the product owner describes the highest priority features to the team. The team asks enough questions that they can turn a high-level user story of the product backlog into the more detailed tasks required to build the features to implement the capabilities defined in each story.

The Sprint Planning & Meeting Schedule consists of the following data fields:

- Identification of the Sprint and associated tasks.
- The sprint duration (in days)
- The start and end dates for each sprint
- The task descriptions
- The priority levels for each task
- The story points associated for each task
- The total number of story points assigned to the sprint.

YOUR COMPANY LOGO

Sprint Planning & Meeting Schedule
Project Name
Version

Sprint/Task	Duration	Start Date	End Date	Task Description	Priority Level (1=Low, 10=High)	Story Points
Sprint 0	99 Days	99-99-99	99-99-99			999 Total
Sprint 0 Planning	9 Days	99-99-99	99-99-99		??	99
Sprint 0 Development	9 Days	99-99-99	99-99-99		??	99
Sprint 0 Review	9 Days	99-99-99	99-99-99		??	99
Sprint 0 Retrospective	9 Days	99-99-99	99-99-99		??	99
Sprint 1	99 Days	99-99-99	99-99-99			999 Total
Sprint 1 Planning	9 Days	99-99-99	99-99-99		??	99
Sprint 1 Development	9 Days	99-99-99	99-99-99		??	99
Sprint 1 Review	9 Days	99-99-99	99-99-99		??	99
Sprint 1 Retrospective	9 Days	99-99-99	99-99-99		??	99
Sprint 2	99 Days	99-99-99	99-99-99			999 Total
Sprint 2 Planning	9 Days	99-99-99	99-99-99		??	99
Sprint 2 Development	9 Days	99-99-99	99-99-99		??	99
Sprint 2 Review	9 Days	99-99-99	99-99-99		??	99
Sprint 2 Retrospective	9 Days	99-99-99	99-99-99		??	99
Sprint 3	99 Days	99-99-99	99-99-99			999 Total
Sprint 3 Planning	9 Days	99-99-99	99-99-99		??	99
Sprint 3 Development	9 Days	99-99-99	99-99-99		??	99
Sprint 3 Review	9 Days	99-99-99	99-99-99		??	99
Sprint 1 Retrospective	9 Days	99-99-99	99-99-99		??	99
Sprint 4	99 Days	99-99-99	99-99-99			999 Total
Sprint 4 Planning	9 Days	99-99-99	99-99-99		??	99
Sprint 4 Development	9 Days	99-99-99	99-99-99		??	99
Sprint 4 Review	9 Days	99-99-99	99-99-99		??	99
Sprint 4 Retrospective	9 Days	99-99-99	99-99-99		??	99
Sprint 5	99 Days	99-99-99	99-99-99			999 Total
Sprint 5 Planning	9 Days	99-99-99	99-99-99		??	99
Sprint 5 Development	9 Days	99-99-99	99-99-99		??	99

www.SDLCforms.com

SPRINT BACKLOG STATUS (WORKSHEET VERSION)

The data included within the *Sprint Backlog Status Worksheet is an output of the* Sprint Planning & Meeting Schedule. The worksheet is used to identify the status of the work that remains on the project. The sprint backlog contains additional details not identified on the Product Backlog; most importantly, the sprint backlog includes work tasks defined by the development team for each of the user stories, and their status in the development process.

The Backlog Status worksheet can be created in a worksheet application, such as MS Excel, or alternatively in a diagramming and vector graphics application, such as MS Visio. Worksheet applications tend to be more available across an organization, while the diagramming and vector graphics application version employs "drag and drop" capabilities to facilitate moving and copying data objects within the document.

The Sprint Backlog Status worksheet can be used to identify only those tasks that are necessary within a single sprint, or across all sprints and releases.

A Sprint Backlog Status Worksheet includes the following data fields:

- Sprint identification
- Product Backlog item (Theme or Epic)
- Tasks (To Do, Development, Verify, Done)
- Owner
- Remaining work in Hours

YOUR COMPANY LOGO

Sprint Backlog Status
Project Name
Version

Sprint	Product Backlog Item	To Do	Development	Verify	Done	Owner	Remaining Work in Hours
99	Epic/User Story	Task			Task	Name	99
	User Story		Task			Name	99
	User Story		Task			Name	99
99	Epic/User Story		Task Task			Name	99
99	Epic/User Story	Task		Task	Task	Name	99
99	Epic/User Story		Task			Name	99
99	Epic/User Story	Task Task			Task Task	Name	99
99	Epic/User Story		Task	Task	Task	Name	99
99	Epic/User Story		Task			Name	99
99	Epic/User Story			Task		Name	99
99	Epic/User Story			Task Task	Task Task	Name	99

www.SDLCforms.com

SPRINT BACKLOG STATUS (GRAPHICS APPLICATION VERSION)

The data included within the *Sprint Backlog Status Worksheet is an output of the* Sprint Planning & Meeting Schedule. The worksheet is used to identify the status of the work that remains on the project. The sprint backlog contains additional details not identified on the Product Backlog; most importantly, the sprint backlog includes work tasks defined by the development team for each of the user stories, and their status in the development process.

The Backlog Status worksheet can be created in a worksheet application, such as MS Excel, or alternatively in a diagramming and vector graphics application, such as MS Visio. Worksheet applications tend to me more available across an organization, while the diagramming and vector graphics application version employs "drag and drop" capabilities to facilitate moving and copying data objects within the document.

The Sprint Backlog Status worksheet can be used to identify only those tasks that are necessary within a single sprint, or across all sprints and releases.

A Sprint Backlog Status Worksheet includes the following data fields:

- Sprint identification
- Product Backlog item (Theme or Epic)
- Tasks (To Do, Development, Verify, Done)
- Owner
- Remaining work in Hours

Sprint Backlog Status
Project Name
Version

Sprint	Product Backlog Item	Tasks				Owner	Remaining Work in Hours
		To Do	Development	Verify	Done		
99	Ep;ic / User Story					Name	99
99	User Story	Task Task				Name	99
	User Story		Task Task Task			Name	99
99	User Story			Task		Name	99
99	User Story				Task	Name	99
99	User Story			Task Task	Task	Name	99
99	User Story	Task	Task	Task	Task Task	Name	99
99	User Story			Task		Name	99

www.SDLCforms.com

14-DAY SPRINT BACKLOG – CAPACITY TEMPLATE

The *14-day Sprint Backlog – Capacity Template* contains the data output from a sprint planning meeting. Specifically, it contains a list of tasks that the Scrum team needs to complete during the sprint in order to convert a selected set of product backlog items into deliverable functionality within a sprint duration period of 14 days.

As you can see in the next section, users can modify the duration specified within the template titled Sprint Backlog - Capacity Template. The duration specifies how many days the team allocates to each Sprint. The 14-day Sprint Backlog identifies the various sprints, and capacity for an elapsed period of 14 days.

The 14-day Sprint Backlog – Capacity Template consists of three sheets: 1) a 14-day sprint backlog table, 2) a capacity sheet, and 3) an instruction sheet providing directions for updating the tables.

Formulas: Formulas are embedded in the chart to calculate:

- Remaining Time for Work Item Each Day = Max (all previous Spent entries).
- Day Totals – total of Spent and Remain columns.

Sprint Backlog Sheet	
Work type	Suggested entries include "Feature," "Tax," "Precondition," and
Deliverable Area	Enter the Deliverable Area.
Product Backlog Item or Iteration	Enter this description.
Work Item ID	Enter the appropriate ID.
Sprint Work Item Description	Enter a description of the work to be accomplished.
Responsibility	Enter who will be responsible for the work.
Status	Suggested entries include "Complete,
Priority	Implement preferred
Initial (Estimate)	Enter the initial estimate for the work item.
Spent	As each day passes, enter the time spent during the day working
Remain	This amount is automatically calculated by the embedded formulas
Day Totals	At the bottom of the spreadsheet, totals for each day will display.

Capacity Sheet	
The initial tasks you need to complete to prepare this sheet for a project meeting are as follows:	
Current Date	Enter the day of the week.
Sprint Start Date	Enter the date in a mm-dd-yy format.
Scrum Team	Enter the names of the individuals on the Scrum Team.
% of Time Allocated to Project	Enter the % of time each individual can devote to the team.
Assigned Hours	Enter the total assigned hours for each individual.
Date	Enter the date for each of the days.
Hours	Enter the number of hours worked each day.

14-Day Sprint Backlog – Capacity

YOUR COMPANY LOGO

14-Day Sprint Backlog-Capacity
Project Name
Version

Worktype	Deliverable Area	Product Backlog Item or Iteration	Work item ID	Sprint Work item Description	Start Date Respons - ibility
					Name
					Name
					Name
					Name
					Name
					Name
					Name
					Name
					Name
					Name
					Name
					Name
					Name
					Name
					Name
					Name
					Name
					Name
					Name
					Name
					Name
					Name
					Name
					Name
					Name
					Name

www.SDLCforms.com

30-DAY SPRINT BACKLOG – CAPACITY TEMPLATE

The *30-day Sprint Backlog* – Capacity Template contains the data output from a sprint planning meeting. Specifically, it contains a list of tasks that the Scrum team needs to complete during the sprint in order to convert a selected set of product backlog items into deliverable functionality within a sprint duration period of 14 days.

Note: as you can see from the previous section the duration specified within the Sprint Backlog - Capacity Template can be modified to support however many days the team decides to allocate to a Sprint. The 30-day Sprint Backlog identifies the various sprints, and capacity for an elapsed period of 30 days.

The 30-day Sprint Backlog – Capacity Template consists of three sheets: 1) a 30-day sprint backlog table, 2) a capacity sheet, and 3) an instruction sheet providing directions for updating the tables.

Formulas: Formulas are embedded in the chart to calculate:

- Remaining Time for Work Item Each Day = Max (all previous Spent entries).
- Day Totals – total of Spent and Remain columns.

Sprint Backlog Sheet	
Work type	Suggested entries include "Feature," "Tax," "Precondition," and
Deliverable Area	Enter the Deliverable Area.
Product Backlog Item or Iteration	Enter this description.
Work Item ID	Enter the appropriate ID.
Sprint Work Item Description	Enter a description of the work to be accomplished.
Responsibility	Enter who will be responsible for the work.
Status	Suggested entries include "Complete,"
Priority	Implement preferred scale for designating Priority.
Initial (Estimate)	Enter the initial estimate for the work item.
Spent	As each day passes, enter the time spent during the day working
Remain	This amount is automatically calculated by the embedded formulas
Day Totals	At the bottom of the spreadsheet, totals for each day will display.

Capacity Sheet	
The initial tasks you need to complete to prepare this sheet for a project meeting are as follows:	
Current Date	Enter the day of the week.
Sprint Start Date	Enter the date in a mm-dd-yy format.
Scrum Team	Enter the names of the individuals on the Scrum Team.
% of Time Allocated to Project	Enter the % of time each individual can devote to the team.
Assigned Hours	Enter the total assigned hours for each individual.
Date	Enter the date for each of the days.
Hours	Enter the number of hours worked each day.

YOUR COMPANY LOGO

30-Day Sprint Backlog-Capacity
Project Name
Version

Worktype	Deliverable Area	Product Backlog Item or Iteration	Work item ID	Sprint Work item Description	Start Date Owner
					Name
					Name
					Name
					Name
					Name
					Name
					Name
					Name
					Name
					Name
					Name
					Name
					Name
					Name
					Name
					Name
					Name
					Name
					Name
					Name
					Name
					Name
					Name
					Name
					Name
					Name

www.SDLCforms.com

PRODUCT BACKLOG GROOMING TEMPLATE

The *Product Backlog Grooming Template* manages data from the product backlog grooming meetings. The product backlog grooming meetings are used to evaluate the themes, epics, and stories and refine the priorities for development going forward.

Product Backlog Grooming is a meeting that is held near the end of each sprint to ensure the backlog is reviewed and refined before committing to work in the next sprint. The intent of a "grooming" meeting is to ensure that the backlog remains populated with specified requirements that are currently relevant, prioritized, complete, and estimated at a high-level, before further refinement of the stories within the Sprint planning meetings.

During a product backlog refinement meeting, the team and product owner discuss the top items on the product backlog and to record their decisions within the Product Backlog Grooming template.

Note: Backlog refinement is really a checkpoint on priorities rather than an effort to fully resolve and development related issues. Development tasks and issues related to each task are discussed and addressed in the Sprint planning meetings.

Tasks performed in the Product grooming meetings include:

- Removing user stories that no longer appear relevant
- Recording new stories in response to newly discovered needs
- Re-assessing the relative priority of recorded stories
- Assigning initial story point estimates, if available
- Correcting story point estimates in light of newly discovered information
- Splitting stories that are too large to fit in an upcoming sprint.

The Product Backlog Grooming Template contains the following data fields:

- Sprint number
- Product Backlog Item description
- Description of issues that are impediments or concerns
- Record the identity of new stories
- Record deletion of stories that are no longer required
- Provide a status of the user story (To Do, Development, Done)
- Identify the stories priority
- Record of allocated story points.

Product Backlog Grooming
Project Name
Version

Sprint	Product Backlog Item	Issues	New Story	Delete Story	Status			Priority	Story Points
					To Do	Develop-ment	Done		
99	Epic/User Story				X			4	99
99	User Story					X		3	99
99			User Story		X			6	99
99	User Story	Is this story really required?		X				8	99
99	User Story						X	7	99

www.SDLCforms.com

ITERATION RETROSPECTIVE TEMPLATE

An Agile or Scrum Retrospective is a meeting that's held at the end of each iteration by team members to assess their performance and seek better approaches to their work in the future. This is similar to defining lessons learned in the traditional model, but much more responsive and proactive in terms of effecting work on the current project.

The team reflects on how everything went and then decides what changes they want to make in the next iteration. The retrospective is accomplished by team collaboration, and team members should decide together how the meetings will be run and how decisions will be made about improvements. In order for this process to work the team must buy into the notion of total transparency and they must not fear the information provided in the retrospective meeting will be used against them. The goal is to improve but not condemn the performance of the team.

During the retrospective, team members are asked the following questions:

- What worked well and therefore should want to do again?
- What did we learn from the iteration/sprint metrics?
- What did we learn that we want to do differently?
- Do we need to refine our assessments on the range of story points story points possible within in iteration?
- What did we learn in the flow of events?
- What can we do better in reporting project status?
- What did we NOT investigate or talk about that likely could stand improvement?

The Iteration Retrospective contains the following data elements:

- Date of the retrospective
- Project manager's name.
- Who completed the retrospective?
- What questions will be asked in the retrospective meeting?
- What worked well?
- What can be improved?
- What do we want to try?
- What do we still not know?

Explore Phase
Overview

In the previous Speculate phase the work focused on capturing requirements, priorities, and supporting critical meetings such as product grooming, sprint planning, and sprint retrospectives. In the explore phase, the attention turns to delivery of work - including the team self-organization to most efficiently address the current workload, encouraging team collaboration, communications, and team interactions with each other, the product owner, customers, users, and other important stakeholders.

Where Speculate phase activities have a focus on planning and defining requirements, the Explore Phase has a focus on action. As discussed in the authors introduction within this module, agile is a set of values and principles, and not prescriptive set of directions on how development work occurs. And also mentioned the signatories of the agile manifesto represented at least 10 different lightweight software development approaches they considered to meet the criteria defined in the agile manifesto. In other words, the software development tasks associated with this phase are highly dependent upon the technologies and related methods employed for development.

As a result, this explore phase only has one agile template associated with it, the story completion checklist. That does not mean to imply there is not a lot of agile-based work going on in this phase. For example, within the scrum methodology, the team participates in short, daily scrum meetings. In those meetings they will address the following three questions:

1. What work was accomplished since we last met?
2. What worked we plan to get accomplished before we meet again?
3. What impediments are in our way of getting the work done?

A template can be created to record the daily scrum meeting. But the meeting approach is simple enough that a simple document, email, or other tool can be used keep track of the daily meeting activities and decisions.

Rather than resolve any questions during the daily scrum meetings, affected members will meet off-line to work through the issues. Also, the scrum master will help them work through any impediments so that the development team members can stay focused on their work.

Other work within this phase includes assessments by the developers or assigned software or systems architects on the architecture and design changes necessary to support the work in the upcoming sprint. The development team leads, and other team members, may also work off-line with customers and end users to further refine the user stories where the understanding is not sufficient to implement as a feature or function within the application.

Forms & Templates

STORY COMPLETION CHECKLIST

As Agile stories are completed, the product owner will work with their Agile team and the team's scrum master to make sure that finished stories meet the *Definition of Done*. This means every task is completed for every story, and that the work products meet acceptance criteria and are ready for review by the product owner.

However, in practical reality, there is more to the stat of being "Done" than this simple definition implies. In order to status a user story as "Done," it must also meet the following criteria:

- Is the story 100% tested?
- Were automated tests created and executed?
- Are all tests checked into the appropriate test case management system?
- Are all product artifacts properly versioned and checked into the Source control management (SCM) system?
- Have all builds/branches been integrated into the mainline of code?
- Have the customer requirements and acceptance criteria been fully met?
- Have customers/stakeholders viewed and accepted the tasks?
- Are changes fully documented in internal specs, etc.?
- Are changes fully reflected in customer-facing documentation (i.e. user manuals, etc.).?
- Are all customer reviewers satisfied with the functionality?
- Are there any remaining caveats or gotchas that could impact release?
- Have negative test been demonstrated as well as positive?
- Is all necessary error checking in place?
- Have interfacing components/ features been tested?
- Are all external dependencies met?
- Are any necessary legal agreements in place to use the work?
- Have workflows been reviewed by usability teams?
- Have text strings been reviewed by legal/UI/usability?
- Are all necessary text string translations in place?
- Are all localization requirements met?
- Have any newly discovered work items been recorded in the backlog?
- Are any bugs that are not going to be fixed documented in release notes/ problem tracking items?
- Has the story been marked as complete in agile project management systems?
- Is the team sure that no other work is necessary?
- Other unique concerns that are specific to the project or product.

The *Story Completion Checklist* is a tool to work through the above questions, to ensure the product is truly done.

Story Completion Checklist
Project Name
Version

Note: In any table, select and delete any blue line text; then click Home→Styles and select "Table Text" to restore the cells to the default value.

Story Name	*Provide customer feedback*	Developers	*Betty Smith, Brad Bards*		
Feature Name	*Develop online form for customer*	Inspected By:	*Joe Jones*	Date:	*99/99/99*

Story Completion Checklist

Legend: C = Completed, I = Inspected, # = Number of defects

Activity	C	I	#	Activity	C	I	#
Story scope and estimate reconsidered	☐	☐	*3*	*Manual tests completed*	☐	☐	
Development environment created	☐	☐	*1*	*Automated tests completed*	☐	☐	
Requirements drafted	☐	☐	*0*	*Test design completed*	☐	☐	
User interface created	☐	☐	*0*	*Code complete*	☐	☐	
Technical design drafted	☐	☐	*0*	*Manual tests executed and passed*	☐	☐	
User documentation drafted	☐	☐	*2*	*Technical design documentation completed*	☐	☐	
Test design drafted	☐	☐	*0*	*User interface documentation completed*	☐	☐	
	☐	☐		*User documentation completed*	☐	☐	
Requirements completed	☐	☐	*2*	*Story integrated (& integration tests passed)*	☐	☐	
Existing manual tests defined	☐	☐	*0*		☐	☐	
Tests drafted	☐	☐	*1*		☐	☐	
Code drafted	☐	☐	*0*		☐	☐	
	☐	☐			☐	☐	
	☐	☐			☐	☐	
	☐	☐			☐	☐	
	☐	☐			☐	☐	
	☐	☐			☐	☐	
	☐	☐			☐	☐	
	☐	☐			☐	☐	
	☐	☐			☐	☐	
	☐	☐			☐	☐	
	☐	☐			☐	☐	
	☐	☐			☐	☐	
	☐	☐			☐	☐	
	☐	☐			☐	☐	
	☐	☐			☐	☐	
	☐	☐			☐	☐	
	☐	☐			☐	☐	

www.SDLCforms.com

This page is intentionally left blank

Adapt Phase
Overview

One of the critical understandings that agile practitioners have is that following a plan suggest that there are best practices that apply to all similar situations. That actually turns out to be seldom true. Best practices, or sometimes referred to as "good practices," are at best starting points or recommendations for how do work. The team must review each story with this notion in mind as they define and refine their work tasks to implement the required capabilities.

From a project management perspective, control and corrections are common terms applied to this phase or type of agile-based work. Plans are made, results are monitored, and corrections are made. As noted in this section on retrospectives, there is no place for judgment or accusations in this place. "Adapt" implies modification or change rather than success or failure. In projects guided by the philosophy - that responding to change is more important than following a plan, attributing failure to variation from the plan isn't productive.

Those who work purely from in ad hoc basis may fail to learn from their mistakes, whereas with those who incorporate and retain lessons learned understand improvements are possible even in highly volatile environments. The agile-based methods and tools associated with the adapt phase a relatively simple compared to the reporting requirements implemented under the traditional model. The focus of agile-based monitoring and reporting is on the team's assessments of work they can accomplish with any Sprint, their performance against those assessments, and ultimately to provide estimates on when the team can realistically complete the work identified in the product backlog.

The adapt phase assessment tools included in this book are:

1. Sprint burn up chart
2. Sprint burn down chart
3. Sprint velocity chart

Forms & Templates

SPRINT BURNUP WORKSHEET AND CHART

A Sprint Burnup worksheet and burnup chart is a graph that shows the progress of work toward a *goal line, usually shown in accumulated story points,* displayed with time values along the horizontal axes and story point values displayed along the vertical axis.

A burn up chart typically includes two lines on the chart:

1. Total work line
2. Work completed line

The total work line depicts the accumulated estimated story points for this total scope of work outlined in the product backlog. The total work values also vary over time as the project owner and development team reassesses changes in requirements and priorities over the course of the project. In contrast, the work completed line shows the accumulated story points to date. So, over time, the work completed line will march upwards and to the right until it matches the total work line. By definition, when that happens, the project work is completed.

In other words, as work is completed over *time* (the horizontal axis), the progress line moves up (burns up) toward the desired total work line. We can also show projected outcomes on the burnup charts, beyond the current date, by calculating a trend line to see when work might be completed.

These charts help identify certain types of problems, such as scope creep or a deviation from the planned project path. The value the Burnup charts is they help identify potential scope and resource problems with sufficient time to take corrective action, or at least not set false expectations on the amount of work that can be completed over the course of the project's duration.

The Sprint Burnup worksheet and chart contains entries which permit the generation of the total work line and were completed line within the associated Burnup graphic. The data fields include:

- **Dates or sprint numbers**
- **Total stories to be completed**
- **Number of stories completed to date**

Data Grid: When implemented within a worksheet tool, such as MS Excel, the data grid is constructed as a table, which means whenever you update the table, by adding or deleting rows or columns, or simply changing the values in the table, the graph is automatically updated.

Note that it's also possible to use estimated hours instead of story points; but that is not best practice.

Sprint Burn-Up Chart
Project Name
Version

YOUR COMPANY LOGO

Dates	Total Stories	Completed Stories
1	325	3
2	327	28
3	335	53
4	340	78
5	340	103
6	340	128
7	340	153
8	341	178
9	341	203
10	341	218
11	342	240
12	342	267
13	345	278
14	345	298
15	367	305
16	378	326
17	378	343
18	386	367
19	392	376
20	400	400

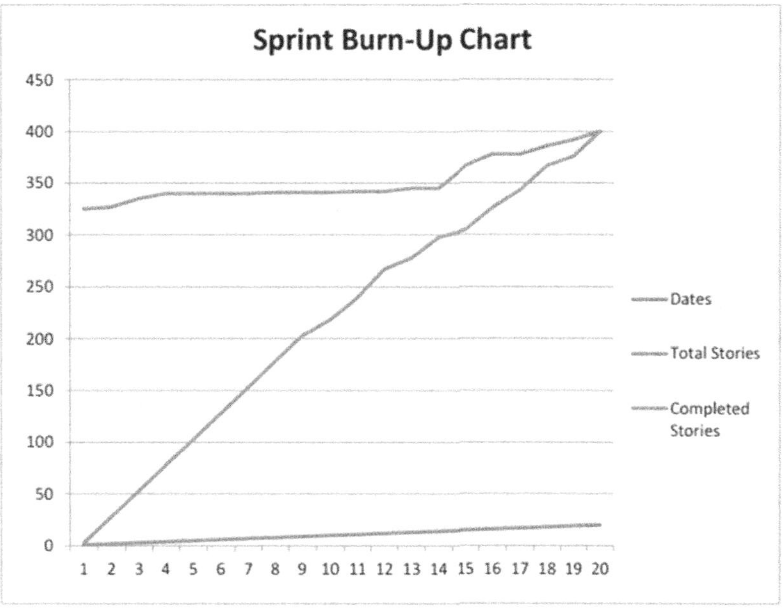

SPRINT BURNDOWN WORKSHEET AND CHART

The Sprint Burndown worksheet and burndown chart has the opposite objective as the Sprint Burnup chart. Instead of depicting how much work is completed and projecting how much time is required to complete the total planned scope of work, the Sprint burndown chart displays the rate at which work is accomplished compared to the estimated story points for each Sprint, and a trendline showing how much work is planned to be completed from the backlog, across each Sprint, from start to finish. So, in other words, the trendline starts with the maximum value of story points with a trend toward zero by the end of the project.

The trend line displays the remaining effort for a given period of time. Work remaining is displayed on the Y (vertical) axis and the number of working days is displayed on the X (horizontal) axis. When the x and y axis values coincide, the work is complete.

When tracking product development using the Burndown Chart, teams can implement both a Sprint Burndown Chart and a Release Burndown Chart. The difference between the two versions of the burn down chart is the focus on the number of story points completed across the scheduled sprints versus the scheduled releases.

The Sprint Burndown worksheet includes the following data fields:

- Product Backlog Item (PBI) e.g., name of each story
- Task number
- Name of the task
- Status (To Do, In-Progress, Closed)
- Original Estimate of time
- An array of 10 days depicting the amount of time spent working on each listed task
    - Note that the net effect of the array is to decrement work effort on a daily basis through the Sprint or release duration.
    - The ideal situation is to complete all work by the last day
- Story points allocated to the sprint review
- Data Grid: The data grid is constructed as a table, which means whenever you update the table by adding or deleting rows or columns or simply changing values in the table, the graph is automatically updated.

Formulas: Formulas are embedded in the worksheet to calculate the story points remaining each day after deducting the previous work effort to date from the maximum story point value. It's also possible to use estimated hours instead of story points; but that is not best practice.

Sprint Burndown Chart
Project Name
Version

Product Backlog Item (PBI)	Task	Name	Status	Original Esimate	Day 1	Day 2	Day 3	Day 4	Day 5	Day 6	Day 7	Day 8	Day 9	Day 10	Sprint Review
PBI Item 1	Task 1	Name1	Closed	7.0	0.0	0.0	3.0	2.0	2.0	0.0	0.0	Day	0.0	0.0	0.0
PBI Item 2	Task 2	Name2	Closed	3.0	0.0	3.0	0.0	0.0	0.0	0.0	0.0	0.0	0.0	0.0	0.0
PBI Item 3	Task 3	Name3	In Progress	4.0	0.0	0.0	0.0	0.0	0.0	0.0	2.0	2.0	0.0	0.0	0.0
PBI Item 4	Task 4	Name4	Closed	1.0	0.0	0.0	0.0	0.0	0.0	0.0	1.0	0.0	0.0	0.0	0.0
PBI Item 5	Task 5	Name5	Closed	0.5	0.0	0.0	0.0	0.0	0.0	0.0	0.0	0.0	0.5	0.0	0.0
PBI Item 6	Task 6	Name6	Closed	0.5	0.0	0.0	0.0	0.0	0.0	0.0	0.5	0.0	0.0	0.0	0.0
PBI Item 7	Task 7	Name7	In Progress	8.0	0.0	0.0	0.0	0.0	0.0	4.0	0.0	2.0	0.0	2.0	2.0
PBI Item 8	Task 8	Name8	Closed	2.0	0.0	0.0	0.0	0.0	2.0	0.0	0.0	2.0	0.0	0.0	0.0
PBI Item 9	Task 9	Name9	Closed	1.0	1.0	0.0	0.0	1.0	0.0	0.0	0.0	0.0	0.0	2.0	0.0
			Estimated Burndown	27.0	24.3	21.6	18.9	16.2	13.5	10.8	8.1	5.4	2.7	0.0	
			Daily Complete	32.0	1.0	3.0	3.0	3.0	4.0	4.0	3.5	6.0	0.5	4.0	2.0
			Actual Burndown	32.0	31.0	28.0	25.0	22.0	18.0	14.0	10.5	4.5	4.0	0.0	

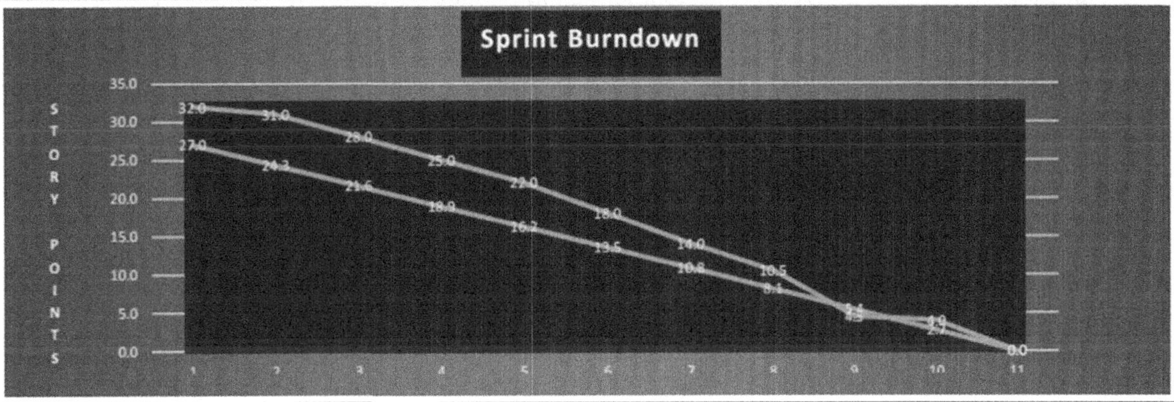

www.SDLCforms.com

SPRINT VELOCITY CHART

In several instances, this book has pointed out that a development team can better estimate the work required to complete development tasks across each user story in the form of story points, that depict the comparative degree of difficulty. The previous Sprint Burndown and Burnup Charts used those story points to indicate progress against the total scope of work defined in the product backlog. But it's also useful to track the scope of work completed within each Sprint in terms of story points actually completed. This is the purpose of the Sprint velocity charts.

In other words, the Velocity Chart shows the amount of value delivered in each sprint, defined in terms of total story points, which provides two important pieces of information.

An average per Sprint number of story points - to better estimate how much work can be done in future sprints

A trendline - showing whether the team is getting better or worse at estimating or accomplishing work

The better the estimates, the better the team can predict and demonstrate their ability to accomplish the total scope of work with the resources and time allotted to the project. This need to demonstrate performance capabilities is no different on agile-based projects than it is for traditional-based projects. A development team is deluding itself if it thinks the executive sponsors or paying customers won't shut down a project that they believe is underperforming, or unable to meet its delivery commitments.

Data Grid: The Velocity Chart data grid is constructed as a table, which means whenever you update the table by adding or deleting rows or columns or simply changing values in the table, the graph is automatically updated.

Formulas are embedded in the chart to calculate the upper and lower charts in the sample image. Sprint iterations are plotted on the horizontal axis. The story points are plotted on the vertical axis. The plotted values include the values within the three columns titled "Iteration/ Sprint," "planned story points," "actual story points."

Sprint Velocity Chart
Project Name
Version

Iteration / Sprint	Planned Story Points	Actual Story Points
0	30	32
1	30	36
2	30	24
3	30	48
4	30	32
5	30	30
6	30	34
7	30	23
8	30	22
9	30	54
10	30	23
11	30	19
12	30	41
13	30	27
14	30	27
15	30	28
16	34	58

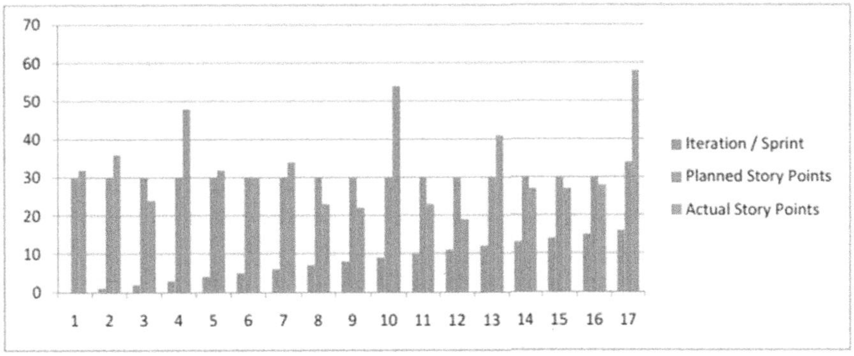

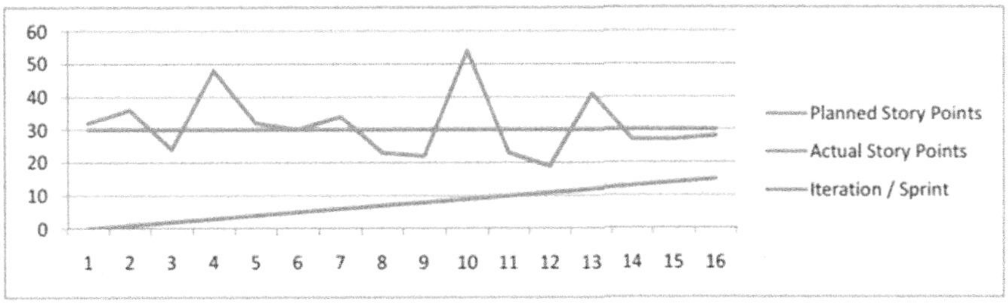

www.SDLCforms.com

Close Phase
Overview

Projects are partially defined as having characteristics that there is a specific beginning and an end. Modern DevOps practices involving continuous development activities, often spanning multiple fiscal years, can blur this distinction. Nevertheless, most projects have definitive expectations in the form of portfolio objectives and deliverables within each fiscal year. Project and development teams need to work with that understanding in mind. Projects have an ending, and that ending, when successful, should also include a celebration of the team's accomplishments.

Note that there are also closing type activities associated with the end of each sprint, and release. Under agile, the team may schedule events, such as the scrum review and retrospective meetings, to capture lessons learned and to discuss ways the team can improve its performance in future iterations.

However, beyond the sprint review and retrospective events noted above, there is little distinction between closing out an agile-based software or systems development project versus a traditional-based software or systems development project. The basic activities remain the same, such as:

- Obtain customer approvals on final deliverables
- document lessons learned for future projects
- Offboarding project team members and contractors. Reassignment of employees to their functional departments or onto other project or program teams
- Closing out any contract agreements related to the project
- Ensuring all configuration items and project artifacts are properly archived in the organization's configuration management and source control management systems
- Complete any final required reporting and payment requirements

Rather than duplicate the methods and templates associated with this agile-based close phase, you will find related methods and tools described in the close phase within Module Two – Traditional Software and Systems Development Model.

Module 02. Traditional Tools and Templates

AUTHOR'S INTRODUCTION

This module provides a description of methods and tools used within the so-called traditional SDLC model, otherwise known as "Waterfall." In my *Frameworks for agility* book I spend quite a bit of time discussing the history of SDLC practices and explaining how agile improves on the traditional SDLC model. But that's not to say that everything in the traditional model is all bad.

From a historical perspective, the traditional model essentially followed project-based management principles, as software and systems developed shared common characteristics with project-based work. For example, software projects are often relatively unique in terms of work and deliverables, and often large and complex endeavors. Project-based work is typically funded under customer directed constraints that dictate the scope of work that is authorized, the budgets allocated to the effort, the delivery schedules, the resources allotted to the project, and desired quality attributes of the completed solution.

So, it seemed logical, initially, the software systems development projects should be managed in the same way that we might manage a project to build a utility facility, or ship, or large building, a road, or a bridge. As it turns, the traditional project-based model created inefficiencies and impediments that got in the way of building good software. Customers want value, timely deliveries of products, and no bugs. The traditional model often failed on all three accounts.

It took the agile movement to address those issues. In the previous module within this book, on agile-based methods and tools, I discussed those issues. To summarize again, the traditional model was plan driven and followed a linear sequential set of processes that delayed the release of products over an extended period of time and made it very difficult to discover and fix the source of bugs and defects in the software when testing is held off until the very end of the development cycle. The traditional model, followed strictly in this manner, often led to unplanned scope creep, delays in the project schedules, increased costs, poor quality, and products that were too often delivered too late to be of value to the customers.

Interestingly, as I noted in my *Frameworks for Agility* book, some of the earliest software development teams, dating back to the mid-1950s, already understood that the traditional project-based development model had to be tweaked to support their efforts. For example, modern agile concepts such as prototyping, iterative development, frequent testing, and the use of tools to simplify software development were finding their way into practice, but on a very limited scale.

It took until the 1990s for the software development community, and the industry as a whole, to explore new approaches to improve their productivity and the value and quality of their products. There were several business drivers that opened the door for these changes. For example, there were way too many failed projects. In addition, software applications increasingly became critical to streamlining business processes. At the same time, globalism and competitive market pressures demanded rapid business response, not only in changes to business operations, but in changes to the systems that supported those business operations.

A number of so-called lightweight software development methodologies were introduced during this timeframe, such as *Adaptive Software Development, Crystal, Dynamic Systems Development Method (DSDM), extreme programming (XP), Scrum*, and others. By the time the agile manifesto was written, in 2001, these lightweight software development methodologies were already beginning to gain acceptance across the software industry.

I noted a few paragraphs back that not everything is bad with the traditional model. In fact, there are instances when the traditional model is still preferred. For example:

- **When executive management needs to constrain the project's budget, resources, schedule, and scope of work**
- **When government mandates and regulations require detailed process controls and documentation**
- **When the customers won't support frequent product reviews or frequent product releases – this may be the number one reason in some organizations.**
- **When the culture of an organization is slow to change and develop new methods and skills**
- **When a project is considered low risk and the requirements are not subject to change over the duration**

Truthfully, with a little time and effort, agile practices can be made to work within any of the above listed situations. But, when the traditional practices work for them, there may be little motivation to make the change.

But there are work practices within the traditional model that still have value, particularly in the areas of documenting engineering practices for architecture, design, development and testing, deployment practices, and configuration management practices. Yes, we can and should eliminate wasteful efforts that had no value to the customer. Plus, many of the engineering practices are highly influenced by the technology used by the organization to develop the applications. With that in mind, the methods and tools documented within this book are malleable and should be tailored to meet the needs of your team and in context with the technology used to build the organization's products.

CHARACTERISTICS OF THE TRADITIONAL SDLC MODEL

You can pretty much read the back half of each stated value within the agile manifesto to understand the primary characteristics of the traditional SDLC model. Rather than recopy the four agile values from the agile manifesto, which Arie did in module one, this time let's take a look at just the parts that agile practices value less.

- Processes and tools
- Comprehensive documentation
- Contract negotiation
- Following a plan

Process and Tools

There is no doubt that there are many more processes and tools identified in the traditional model than virtually all of the agile-based processes. By definition, the original agile-based SDLC models were considered lightweight. From a practical standpoint the whole point of the traditional model was to control risks by developing rigid processes and best practices, and then using extensive documentation to essentially verify the work was performed in accordance with the defined processes.

Processes are higher level descriptions of work that often include any number of lower-level sub processes, activities, and tasks. Not doing the work correctly is another example of non-value-added effort. Why would a customer want to pay for work that is not efficient or does not produce a quality product? So, under process improvement and maturity programs, organizations document reusable methods that represent good or best practices. And, they may also procure or develop and use tools that help the practitioners perform the work efficiently and correctly.

On the negative side, the costs and effort required to implement rigorous processes, methods, and tools is all overhead, most of which is not value-added. On the positive side, having sound work methods and tools can improve efficiencies, consistency, and quality of work effort across any number of projects over time. The organization must find balance when determining how much effort they should put into developing their SDLC processes.

Comprehensive Documentation

As noted above, documentation under the traditional model helps ensure compliance with following the organizations prescribed processes. But a customer doesn't want to pay for documentation if it is not adding value to building the right products. On the other hand, some documentation is critical to the long-term sustainability of a product.

One example is configuration management documentation that identifies the version of code and other artifacts associated with each release of a product. Developers and contractors will come and go over the life cycle of a product, and when they do leave a project, without proper configuration management, with them goes the knowledge of how the applications were built, and the requirements that drove development of the features and functions within the product.

There are also training, supportability, and maintenance requirements that need some degree of documentation to sustain the information across the product's life cycle. When legacy systems need to be enhanced, it's important that architecture, design, code, test scripts, test results are properly maintained related to each surviving product release.

As noted within the above section on processes and tools, the amount of documentation required to support a product across its life cycle is similarly a matter of balance. Too much is not value-added, and too little will drive up the sustainability costs over the lifecycle of a product.

Contract Negotiations

Whether specified under a project charter or formal business agreement, contract negotiations imply there is a procedure in which the customers and the development team come to terms with defining the scope of work, budgets, and schedules. The connotation is a negative one as it implies there is not shared interests between the customers and developers. Each is trying to get the best deal, which implies there is a cost of some sort that the other party must agree to.

In actual fact, the more likely circumstance is that the customer dictates the constraints of the project leaving the development team, if they are contractors, to decide whether they want the job. If the development team consists of internal employees of the company, they have even less choice about taking on a product, short of leaving the company. In either case, the negotiated approach does not lead to a collaborative environment between the customer and the developers.

From a practical point of view, having managed numerous software development projects, the bigger issue with contract negotiations and formal project charters is that too often the customers believe they have given you their requirements sufficient to build the product they want. Nothing could be further from the truth. What they more likely have given the development team is simply a description of the high-level deliverable product. The devil is always in the details, and those details can only be worked out through a collaborative trial and error process where customers and developers work through the requirements as implemented features.

Following a Plan

The linear sequential approach of the traditional software development model is what's at issue here. The plan includes descriptions and schedules of the processes, activities and tasks necessary to complete the full scope of work. But it's the extension of that work across a lengthy and inflexible development cycle, while attempting to define and build a product all at once, that made it difficult to respond to new customer changes.

I submit that every software development project has some level of planning, whether the projects were managed under agile or traditional SDLC practices. However, the traditional model typically involves detailed initial planning that fixes in scope and schedules. Nevertheless, there is no reason why a project plan cannot provide flexibility to incrementally build a product in smaller parts, providing more opportunities to align the development of features with emergent customer needs and changing priorities.

I mentioned early in this module that some software developers, as far back as 1950s, understood the value of practices common in today's modern agile methodologies. For example, they understood that prototyping and iterative development techniques, with frequent customer reviews, were instrumental to building products customers wanted and delivering in a timely manner. However, these early developers also understood that having sound engineering practices was fundamental and crucial to their overall success.

Certainly, modern software engineering practices help support the values and principles stated in the agile manifesto. But what I think organizations and individuals need to be careful about forgetting the engineering part. And that is where I think the traditional SDLC model excels. E.g., developing, documenting, improving, and sustaining good engineering and business practices, especially at the large enterprise level. But once again, the organization must balance how much effort they should put into these activities, before the work no longer becomes a value-added activity.

ORGANIZING WORK UNDER THE TRADITIONAL SDLC MODEL

Unlike the agile-based SDLC model, there have been numerous attempts to define standards for naming processes and describing the flow of work within the traditional SDLC model. Probably the best known is the ISO/IEC/IEEE 12207 Systems and software engineering – Software life cycle processes.

As a side note, the latest versions of the ISO/IEC/IEEE 12207 Systems and software engineering standards to not attempt to define a life cycle. In other words, they dodged the bullet to declare whether thy promote agile or traditional life cycle development strategies.

Instead, the standards focus on defining common software and systems engineering and management processes apart from whatever lifecycle development strategy their customers might prefer. Specifically, the ISO/IEC/IEEE 12207:2017 standards organize their identified software and systems engineering processes into two major categories, one for technical management processes, and the other for technical processes.

If you have read my *Frameworks for Agility* book, then you know that I'm a fan of the hybrid SDLC Frameworks. A hybrid SDLC model implements a similar strategy to organize systems engineering and related management processes by types of work, and under a rigid life cycle development model. That way, development teams can pick and choose the management and technical processes that best support their client's needs and the lifecycle development strategy they choose to follow.

Nevertheless, I have not personally seen many organizations strictly adhere to the ISO/IEC/IEEE 12207:2017 standards. In my experience, the ISO/IEC/IEEE 12207:2017 standards often serve as references materials when the organization needs more information on how to execute a certain type of software or systems engineering or related technical management process.

Moreover, across the commercial and federal agencies where I've worked, the tendency is to define SDLC processes uniquely to support their business needs. This just makes sense. If a company does things exactly like their competitors, how do they create a competitive advantage.

On the other hand, there is much more in common from one approach to another than there are differences. The software development organizations may choose to build SDLC processes of their own definition, perhaps as part of a process quality improvement and maturity program. They may also choose to augment their processes with information from one or more industry standards.

However, the real gains in productivity and value come from the end-to-end execution of SDLC processes with minimal non-value-added activities, iterative development practices, frequent releases, and a focus on the customer that allows the team to be more responsive to changing business needs and priorities. Having said that, the organization still has to invest in building skills, methods and tools, around good engineering and technical management practices, that support their software systems development activities.

The organization can hope their engineers are getting this type of software engineering process information from the colleges and technical schools they attend. That does not take away the need for the organization to define the software engineering and technical management processes they prefer, develop a lean and customer centric culture, and then develop and improve on their lifecycle development strategies.

You might be wondering why I've spent so much time on this subject at the start of a module on traditional SDLC based methods and tools. The reason is the traditional SDLC models do a better job of defining, classifying and organizing software and systems engineering related work. In contrast, the agile-based SDLC models typically focus more on implementing software development strategies that achieve the values and principles outlined in the agile manifesto, and not so much on describing specific software and systems engineering related tasks.

At a high-level the two models appear to have similar work tasks. For example, if you look at the work tasks listed inside an agile-based software development iteration, the work looks very similar to what you see in the traditional life cycle model. The primary difference is that the agile models show those work patterns occurring in the repetitive and circular fashion, where the traditional model looks more like a single set of cascading steps or waterfalls.

To further make this point, the basic iterative work patters for Agile were displayed in *Figure 2. Basic Agile/ Iterative Development Workflow Diagram*, while the basic approach to software development under the traditional SDLC model is depicted in figure 1, overleaf.

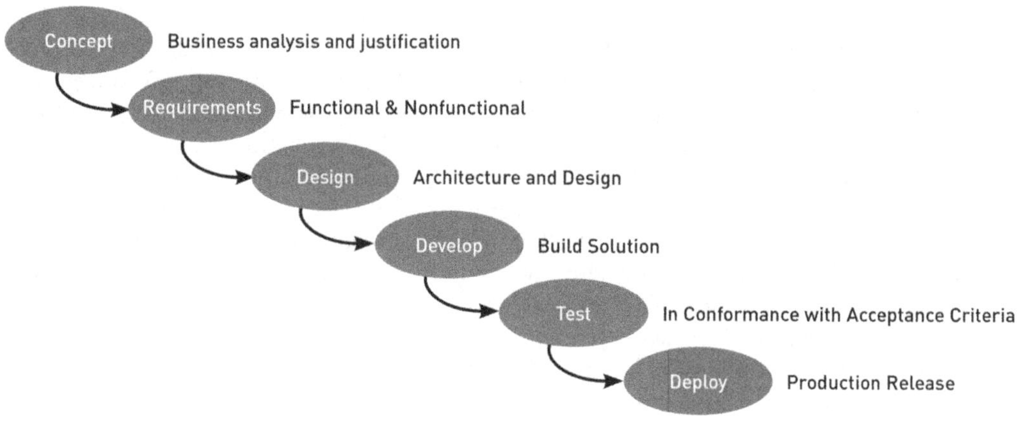

Figure 2, Traditional SLC Model (A.K.A. Waterfall)

You should see that the work patters for both the agile iterations and the traditional waterfall model are essentially the same. The primary difference is agile breaks up the total scope of work into smaller, repetitive, and therefore more frequent development and delivery cycles.

Since our ultimate goal in this book is to improve our software engineering practices through the employment of reusable methods and tools, and since many of the work tasks are similar in both models, we can define an SDLC process organization and classification structure that supports both the agile and traditional models. The easiest way to do that is to simply take the work categories defined in either figure 1 (agile model) or figure 2 (traditional model), and simply organize work into a taxonomy that looks similar to a file directory structure, as shown in figure 4 - *SDLC Taxonomy in Directory Folder Structure*. In fact, that is how the knowledge categories are typically implemented within an SDLC process management tool or portal.

Project and Technical Management Processes.

Across the life cycle of a product there can be any number of projects established and funded to perform work in the initial development, for any modifications and enhancements. It's also not unusual to see individual projects set up to conduct a concept of operations investigation, or business requirements analysis, or deployment related activities. The processes to manage projects are different than the processes to build products, and the management processes occur in parallel across the scope of work authorized for each project.

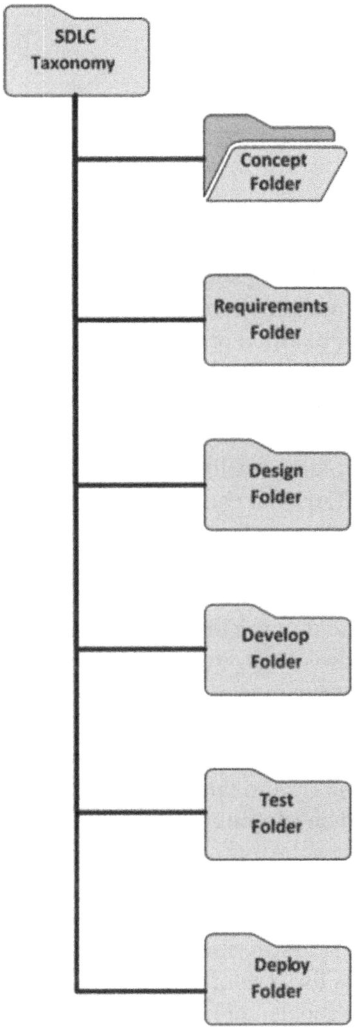

Figure 3. SDLC Taxonomy in Directory Folder Structure

The methods and tools listed within this module are organized in phases that span both the traditional SDLC processes and the traditional project and technology management processes. This is done intentionally as both sets of activities go on in parallel across the life cycle of a product.

The typical project management processes include project initiation, project planning, project monitoring and control, and project close. Technology management related process categories might include decision management, risk management, configuration management, and quality management.

Within each of these project and technical management processes are lower level sub processes, activities, and tasks.

For this book, methods and tools related to project management processes are organized and managed within the following sections:

- **Project Concept/ Initiation Phase**
- **Project Planning Phase**
- **Project Monitoring and Control Phase**
- **Project Close/ Maintenance Phase**

Project Concept Phase
Overview

The project concept phase provides the high-level that is necessary to determine if a business case can be made to initiate a project, whether it is cost justified, and the basic capabilities that must be implemented within the product. The methods and tools supporting development of the project charter, business case, feasibility study, and cost-benefit analysis guide the fact-finding work that is necessary to determine the feasibility of planned programs and projects.

Corporate strategies and mission statements are developed by the chief executives to align organizational resources to better and more efficiently serve the needs of their target market customers. The strategic business assessments focus on identifying business risks and market opportunities. However, strategic assessments lack the implementation details necessary to accurately assess the costs or practicality of developing solutions that can help the organization achieve the goals and objectives of the corporate strategies.

This concept is important to understand. The analysis and data necessary to build a business case, conduct feasibility studies, determine return on investments, and determine high-level capability requirements, follow in response to the development of the organization's executive strategies and mission statements. In other words, the executives need to determine any changes to organizational focus, to remain competitive and build new business, before any work can be done to scope the work necessary to support the stated business objectives.

The concept phase is where the business case, feasibility studies, high level business and user requirements analysis, and cost-benefit analysis occurs. By and large this work has to be supported by the product owner, business analysts, and lead development team members who can drive down on the requirements and more accurately assess the total scope of work required to create a working solution.

Ideally, concept phase-oriented work should be conducted in the fiscal year preceding funding and approval of the actual project. The organization has essentially two paths forward when the concept phase work cannot be completed in the previous fiscal year. One, they can go right into project initiation and hope they can define and develop the desired solutions on time and on budget. Or, two, they can perform the concept phase-oriented work at the beginning of the new fiscal year and understand there will be delays before the project teams can start the work. Both of those options have risks and are not ideal. It's better to get a head-start in the preceding fiscal year.

The primary deliverable of this phase is the Business Case, which includes the information gained from ROI analysis, cost benefit analysis, and feasibility study. But, another important augment to a business case is the Value Proposition. The value proposition helps refine the solution description, the target or niche market, the needs or pains of the target market clients, the client's desired solution capabilities, expected benefits, and whether it's better to make or buy the solution's components.

If there is time and funds available, I also recommend a concept of operations (CONOP) investigation or at least high-level business requirements analysis to better refine the total scope of work. The high-level requirements analysis work helps the development team better understand the technical issues, skills, and resources, and to estimate the time required to build the desired solution.

Not every organization will have a Concept Phase; but it's a good idea - especially on large projects where Waterfall-based SDLC methodologies are contemplated. This is because the organizations that use the traditional linear-sequential SDLC approaches must have a thorough understanding of requirements and the business case that justifies the investments prior to starting development. Under the linear sequential development model, it's impractical to make major changes to the original plans and functionality until after the entire product development cycle is complete.

This is not to imply that all risks are removed by including a concept phase. To the contrary, the very nature of project-based work, having the characteristics of being unique endeavors, means there will always be some degree of risk. What we need to do is manage those risks as best as possible.

Also, until a detailed functional and technical requirements analysis has been completed, during the Requirements and Design Phases, it is truly difficult to assess the total scope of work that may be involved in any given project. Still, the concept phase of work significantly reduces the risks of project failures due to improper scoping.

The deliverables of the concept phase, as described in the previous paragraphs, help ensure that everyone involved with the project understands the parameters under which the project can be judged a success. Also, from a practical perspective, no executive sponsor or customer is going to open up their checkbooks without some sort of limits imposed in terms of budget, time and resources, and expectations set on the deliverable that are required to make the project a success.

Since the paying customer will impose constraints on the project, it's best they do so from a position of knowledge. This statement is just as true for agile-based projects as it is for the traditional software development model.

One last note on this topic. As you can probably tell, I'm not a fan of heuristics at the project level as the means to gauge the scope of work. E.g., "we did a project that looks like this before, so this new project should take roughly the same types of resources and time, and cost about the same." Likewise, corporate executives must not dictate constraints without organizational buy-in.

If corporate or line-of-business executives, or even the project manager or product owner, make scoping and constraint determinations, the project is going to be in trouble from the start. The organization's executives must take the time, and allocate sufficient resources, to properly scope the work and obtain organizational buy-in from all affected project stakeholders and participants.

Forms & Templates

PROJECT CHARTER TEMPLATE

The Project Charter states the business goals and objectives that justified the investments in developing software and system-oriented solutions as an approved project. The charter establishes the constraints on the project, including the authorized scope of work, budget, time, resources and quality objectives. The project's charter also identifies the main stakeholders, and the project and product managers/ product owners.

The Project Charter provides a high level single-page overview of a project for review by executive management. The document ensures that everyone involved in the project is aware of its purpose and objectives.

The Project Charter contains the following topics:

- **Business Case**
- **Opportunity Statement**
- **Business Goal(s)**
- **Project Scope**
  - In-scope processes
  - Out-of-scope processes
- **Project Team**
- **Stakeholders**
- **Project Milestones (by yearly quarter)**
- **Projected costs.**

Note: It's vital to understand the distinction between the word's *product and project.* The product is what's being built as a solution to address business or user needs. Products have a life cycle that can span many years, and products may undergo any number of changes through those years. Projects are the way work is managed over the life of the product, and there may be any number of projects initiated to help define, build, maintain, and support the product over its life cycle. Each project must have a project charter, or some other agreement, that specifies both the objectives and constraints for the new scope of work.

YOUR COMPANY LOGO

Project Charter
Project Name
Version

Business Case

The purpose of this business case is to identify whether there is potential business value to the proposed project or program idea or initiative to commit time, resources, and expenditures. This document provides solution benefits and costs to obtain management approval and secure funding.

Opportunity Statement

Describe in detail the business problem or opportunity addressed by the project or program.

- *Include the business drivers, key systems impacted, and the programs, projects, applications, and departments that are impacted.*
- *Identify the way the project fits with business and corporate strategic goals.*
- *Identify the risk(s) of not doing the project or program.*

Business Goals

List and describe the goals of the project, for example:

- *To reduce costs*
- *To increase efficiency.*
- *To satisfy regulatory requirements*
- *To comply with IT Governance mandates*
- *To satisfy internal/external auditors.*

Project Scope

Provide a high level of the processes/systems that will be impacted by the project, and then identify the in-scope and out-of-scope processes/systems.

In-Scope Processes

Enter the names of the business/IT processes that will be included in the scope of work.

Out-Of-Processes

Enter any processes that are not an integral part of this project.

Project Team

Name	Responsibility
John James	Director, IT
Joe Jones	Manager, IT Governance
Betty Blue	Director, Finance
Peter Panel	Director, Accounting
Robert Hassan	IT Lead

Stakeholders

Name	Responsibility
Ida Grant	Accounting Supervisor
Paul Smith	Finance Manager
Kristine Casey	IT Audit Chief

Project Milestones

	Year			
Milestone	Q1	Q2	Q3	Q4
Analysis				
Requirements Definition				
Development				
Testing				
Production				

Projected Costs

☐	$2,500,000-$5,000,000	☐	$250,000-$500,000
☐	$1,000,000-$2,500,000	☐	$100,000-$250,000
☐	$500,000-$1,000,000	☐	>$100,000

 Document: 2010 Page 1 of 1

www.SDLCforms.com

BUSINESS CASE DOCUMENT

The Business Case Document identifies whether there is a potential business value to justify the investments in developing new solutions as part of the organization's portfolio, program, and project initiatives. Business case must be developed before the organization commits time, resources, and expenditures to the effort.

The Business Case Document opens with a detailed description of the business need, issue or problem and how it will be addressed and supported. As a good practice, the description of the identified business needs should map back directly to the current corporate strategies and mission statement. This helps everyone involved in developing the business case stay focused on staying in alignment with the corporate strategy.

Information within the business case includes descriptions of critical deliverables, quantitative and qualitative measures describing what good looks like for each of the deliverables, identified risks, identified customer and user requirements, procurement issues, constraints on and schedules, plus social, and environmental issues that could impact the success of the proposed solution. The business case also includes information on expected financial gains, cost savings, potential market share, increased usability, and productivity enhancements that would help to justify program or project approvals.

The Business Case, together with the Project Charter and value propositions, provide the information and analysis executive management needs to determine whether a proposed project-based solution development effort supports the mission and strategy of the business, meets target market customer needs, and is justified on an ROI basis.

The Business Case Document contains the following topics:

- **General Information**
  - Request Number
  - IT Sponsor
  - Stakeholders
  - Purpose
  - Project Description
  - Benefits
- **Recommendations and Alternatives**
  - Recommendations
  - Options/Alternatives
  - Critical Success Factors
  - Additional Comments
- **Costs and Resources**
  - High Level Cost Estimate
  - Related Projects
  - Resources
  - Key Milestones
- **Authorization to proceed by stakeholders.**

Business Case Document
Project Name
Version

3 Costs and Resources

3.1 High Level Cost Estimate

- ☐ < $100,000
- ☐ $100,001 - 200,000
- ☐ $200,001 - 500,000
- ☐ $500,001 - 1,000,000
- ☐ $1,000,001 – 3,000,000
- ☐ > $3,000,000.

Note: This high-level estimate is based on the limited information available at the time of the estimate. This estimate can change an average of +/- 50% through the completion of the Business Requirements Document.

Cost Description	Estimated Amount	Capital or Expense
Startup Costs		
Application Costs		
Software		
Hardware		
Implementation		
Restructuring Costs		
Training Costs		
Maintenance		
Support		
Ongoing		
Other		
Total Estimated Costs		

FEASIBILITY STUDY

The Feasibility Study contains business and technical information and cost data to determine the economic potential and practicality (i.e., feasibility) of a proposed project. A feasibility study implements techniques to identify and assess the risks of a proposed project, including alternative approaches.

Business Analysts conducting a feasibility study assess business and technical information, projected costs and benefits, risks, issues and constraints to determine if the project is practical from an economic and technical perspective. In addition, where alternative solutions are available, the feasibility study will determine which proposed solution best fits the needs of the organization with the lowest risks, and documents recommendations made by the analysts.

The Feasibility Study helps to answer various questions that could apply to a prospective project or study:

- How difficult will it be to build and whether the firm has enough experience using a specific technology, and whether the new system will perform adequately?
- How much time is available to build the new system, when can it be built, whether it interferes with normal business operations, type and amount of resources required, etc.?
- What are the contingencies and mitigation plans if the project takes too long to implement or does not work?
- Are there any environmental factors that need to be considered or will the company's own culture clash with the results of the project?
- Are there sufficient human resources available for project development and after it goes live? Do current work practices and procedures support a new system?
- Will the cost/benefit analysis determine the true expected benefits and savings and lead to the next step; a decision to design and implement the system?

The Feasibility Study contains the following topics:

- Introduction, Purpose, Objectives, and Scope
- System Overview and Background
  - System, Plant or Operational Details
  - Current Systems and Processes
  - Current
  - Physical Environment
  - User Organization.
- Deliverables and End Products
- Analysis, Solutions and Alternatives
- Approvals.

Feasibility Study
Project Name
Version

2 System Overview and Background

This section provides an overview and background to the system, plant, or operations, including current procedures and end products.

2.1 System, Plant or Operational Details

Provide an overview and background to the system, plant, or operations, including the following details:

- *Project Owner*
- *Department responsible*
- *System, plant, or operations name*
- *System category, e.g., Hardware / Software / Network, etc.*
- *Operational status, e.g., operational, in development, undergoing modification, special conditions.*

2.2 Current Systems and Processes

Describe the current systems, plants, or operations, and associated business processes.

2.2.1 Current Operations

Describe the current operations. Provide a graphical representation if appropriate.

2.2.2 Physical Environment

Identify the hardware and software technical environment used by the existing system, plant, and/or business processes. Provide a graphical representation if appropriate.

VALUE PROPOSITION TEMPLATE

The value proposition template this similarly one of the best methods to assess whether there is a compelling business case to justify an investment in the development of a new or enhanced solution. Value proposition doesn't replace the business case, rather it augments it, and provides a quick summary of the following information:

- Proposed solution description
- Target market customers
- Descriptions of desired customer capabilities
- Benefits to the customer
- The competition
- Our solution's unique advantages

The value proposition assessment helps an organization determine if there is sufficient value in a proposed software or system solution, from the customer's perspective, to justify moving forward with the project. Later, after the products are developed, the value proposition becomes the basis for the organization's marketing and sales campaigns and communications. In effect, everything a customer needs to know, as to why they should buy our proposed solution offering, is summarized in the value proposition.

The Value Proposition Template contains the following fields:

- Name - of the proposed product and/ or service offerings
- Product/ Project description – proposed scope of work
- Target market – our customers; which can be a niche market within an existing market
- Pains Addressed – customer needs addressed by the proposed solution offering
- Capabilities - needed by our target market customers, and implemented by our proposed solution
- Required features – that implement the desired capabilities
- Benefits – our customers can expect from using our proposed solution
- Competitive Advantage – statements of why customers should prefer buying our products over building their own solution, or purchasing another product from one of our competitors

YOUR COMPANY LOGO

Value Proposition Template
Project Name
Version

Completing the Value Proposition Template will assist an individual/department determine if there is value in a proposed application, system or product, often provided by an outside vendor or contractor, and lend credence to the final decision making process. This template is used in conjunction with the Business Case Document.

Note: In any table, select and delete any blue line text; then click Home→Styles and select "Table Text" to restore the cells to the default value.

Value Proposition Template

Product / Service Offering
Provide the product name and version.
Project Description
Provide a high level description of the customer's problem and how the proposed solution addresses that problem.
Target Market
Who within the organization has relevant pain that is not being addressed? Which in-house organization will most benefit from this solution?
What is Our Pain?
Has the business unit(s) that might benefit from this solution allocated sufficient budget/funds to undertake this initiative?
Capabilities Needed?
What capabilities has the customer identified? What is the customer able to do well, or not do well?
Required Features
Describe the features and functions and how each component will provide the capabilities desired by the customer.
Benefits
Explain how the customer will benefit from the proposed solution in terms of performance, productivity, efficiencies, and cost/ROI.

Make vs. Buy Decisions	
Why is the proposed product superior to in-house development?	*What unique skills, methods, tools, resources, etc., make the proposed product superior to other options?*
What are the advantages of this product over other potential vendors?	

Confidential – ©2015 Documentation Consultants (www.SDLCforms.com) Document: 1150 Page 1 of 1

www.SDLCforms.com

PROJECT COST BENEFIT ANALYSIS TEMPLATE

The Project Cost-Benefit Analysis template facilitates the process of identifying and documenting the proposed project, opportunities, business goals, project scope and issues, alternatives or recommendations, and authorization by key stakeholders. Information gathered from this analysis is incorporated within the value proposition and the larger business case.

The form identifies whether there is potential business value (e.g., summary of analysis on development costs, financial risks, benefits) to commit the time, resources and funding necessary to develop the proposed solution.

The Project/Cost-Benefit Analysis contains the following topics:

- **General Information**
  - Project Name
  - IT Sponsors
  - Stakeholders
  - Purpose
  - Project Description
  - Benefits
  - Assumptions and Constraints
  - Related Projects
- **Recommendations and Alternatives**
  - Recommendations
  - Options/Alternatives
  - Critical Success Factors
- **Costs and Resources**
  - Key Milestones
  - Resources
  - Schedule/Milestone Costs
  - Risks and Impact
  - Risk Analysis
  - Cost/Benefit Analysis
- **Authorization.**

YOUR COMPANY LOGO

Project Cost/Benefit Analysis
Project Name
Version

Costs and Resources

Key Milestones	*Provide a high level schedule or key milestones with dates.*

Development Costs

Estimate the costs of the various phases.

Phase	Year # 1	Year # 2	Year # 3	Total
Concept / Initiation Phase	$	$	$	$
Planning Phase	$	$	$	$
System Requirements Phase	$	$	$	$
System Design Phase	$	$	$	$
Testing Phase	$	$	$	$
Project Monitoring and Control Phase	$	$	$	$
Production Turnover Phase	$	$	$	$
Project Closure / Maintenance Phase	$	$	$	$
			Total:	$

Operational Costs

Estimate the costs of installation, operation and maintenance.

Phase	Start Date	End Date	Total
Personnel			$
Outside Resources			$
System Requirements Phase			$
COTS (Commercial Software)			$
Infrastructure			$
Facilities			$
Supplies			$
		Total:	$

Confidential – ©2015 Documentation Consultants (www.SDLCforms.com) Document: 1250 Page 7 of 9

www.SDLCforms.com

CONCEPT OF OPERATIONS (CONOPS) TEMPLATE

The Concept of Operations, or CONOPS, is a Capabilities Needs Assessment investigation to obtain input from Users' and Stakeholders' on changes that need to be made to a business process or system, or the development of a new system. CONOPS investigation are initiated in advance of the implementation of a strategic change initiative or change in the organization's mission. In other words, the CONOPS is both an analysis and a formal document that describes high-level capabilities requirements that have been identified as necessary to achieve the mission of the business.

Again, it's always important to resist the temptation to move directly to specifying features and functions. Customers have pains or other needs, and they desire to have the capability to resolve those pains and needs.

Product owners and development teams must always address capability needs before determining implementation strategies in the form of features and functions. Therefore, the CONOPS should have a focus on necessary changes to business processes and stay away from software or system implementation details.

The CONOPS captures high-level requirements but is also used as a communications document when making the business case to pursue the necessary investments. The CONOPS provides inputs to a Request for New Applications when the executive sponsors for paying customer requires those documents for their records. The con ops can be used in lieu of completing a Business Requirements Documents (BRD).

The Business case, CONOPS, and BRD all contain similar information but have different purposes. The business case is used to justify financial investments, the CONOPS evaluates capabilities from the perspective of mission support - and typically follows an "As-is/ To-be" format, while the BRD typically provides a deeper dive into the business requirements the solution must implement.

The CONOPS template provides the following sections:

- Purpose
- Capabilities needed
- Operation and Support Description
- Justification for and Nature of Change
- Potential Impacts
- Concepts for a Proposed System
- Scenarios
- Functional Capabilities
- Summary of Impacts
- Analysis of the Proposed System.
- CONOPS Team

YOUR COMPANY LOGO

Concept of Operation (CONOPS)
Project Name
Version

1.6 Purpose of the Proposed System

This sub-section briefly states the purpose of the proposed system or subsystem to which the CONOPS applies. It describes the general nature of the system, and identifies the project sponsors, user agencies, development organizations, support agencies, certifiers or certifying bodies, and the operating centers or sites that will run the system. It also identifies other documents relevant to the present or proposed system.

A graphical overview of the system is strongly recommended. This can be in the form of a context diagram, a top-level object diagram, or some other type of diagram that depicts the system and its environment.

Documents that might be cited include, but are not limited to: the project authorization, relevant technical documentation, significant correspondence, documents concerning related projects, risk analysis reports, and feasibility studies.

1.7 Reference Documents

This sub-section lists the document number, title, revision, and date of all documents referenced in the CONOPS document. This section should also identify the source for all documents not available through normal channels.

Document Number	Title	Source (if applicable)	Revision	Date

1.8 Revision Summary

This sub-section provides a bulleted, high-level description of changes made to the previous version and why. For each revision discussed, provide the date that the revision was made. If the current version in production is the first version of the CONOPS, this page should be left blank below the title.

 Document: 1500 Page 9 of 33

This page is intentionally left blank

Project Start Phase
Overview

The purpose of the project start phase is to ensure completion of project planning work, as appropriate to the proposed scope of the work. For example, the amount of project planning required to support the development or enhancements to a product website is likely to be substantially less than project planning requirements to support implementation of the major enterprise system. The project manager working with the executive sponsors and development team must analyze requirements the project and build a project plan that has just enough guidance while not being overly constrictive or creating unnecessary overhead.

The project start phase includes the methods and templates necessary to initiate a project once the project charter is approved and released. The project plan is the primary artifact associated with this phase of work. The other templates included in this phase can be managed separately or incorporated within the project plan.

There is usually a bit of an overlap between concept and project start phase activities, or at least there should be. The practical reason for the overlap is to ensure the project manager, product owner, and at least the leads on the team development team are involved in the solution analysis and justification activities within the concept phase. If they are not, unwanted and undesirable risk is unnecessarily introduced into the project.

Once the concept phase activities are complete, and sponsoring executives have reviewed the business case, and assuming the executives are inclined to authorize the work, the charter is approved, and the project manager begins to hire and onboard the project team. The project initiation and onboarding checklists aid in this work to ensure nothing is missed from an initial project planning, legal, or human resource perspective.

Other project initiation planning tools include the project or issue submission forms, project team definition form, stakeholder identification list, the project resource plan, and the project management plan template.

PROJECT INITIATION AGENDA

The project initiation agenda template details the agenda for the project's initial "kick-off" meeting. The kickoff meeting should include the executive sponsors, product owner, affected business process owners, key stakeholders - especially any whose opinions could impact the project, affected business managers, and project team members. The agendas distributed to these individuals is sent in advance of the meeting to give participants sufficient time to coordinate their schedules and prepare for the meeting.

The kickoff meeting often serves as the first *meet and greet* opportunity between the team members and their prospective customers. It's important that the project starts off on good footing, and that the participants in the kickoff meeting establish a sense of common purpose and objectives.

The project initiation agenda template starts off by identifying discussion topics and who on the project team will present detailed information on each topic. The remainder of the template contains the following placeholder topics for discussion:

- Meeting Subject
- Meeting Originator
- Meeting Date & Time
- Attendees
- Documents To Read
- Items Required
- Attachments
- Agenda Topic
- Detailed Information
- Presenter.

YOUR COMPANY LOGO

Project Initiation Agenda
Project Name

Text displayed in blue italics is included to provide guidance to the author and should be deleted before publishing the document. In any table, select and delete any blue line text; then click Home→Styles and select "Table Text" to restore the cells to the default value.

Meeting Subject	*Kick Off Meeting / Initiation of Project (Name of Project)*			
Meeting Originator				
Meeting Date / Time				
Attendees				
Documents To Read				
Items To Bring				
Attachments				

Agenda Topic	Detailed Information	Presenter
Team Introduction		
Identify Key Stakeholders and Sponsors		
Review Project Request		
Review Project Process		
Identify Key Business and Technology Sign-off Members		
Project Site Location		
Schedule Business Case or Project Charter Meeting		

Additional Instructions:

(Use this section for additional instructions, comments, or directions)

Confidential – ©2015 Documentation Consultants (www.SDLCforms.com) Document: 1010 Page 1 of 1

www.SDLCforms.com

PROJECT OR ISSUE SUBMISSION FORM

The Project or Issue Submission Form is a one-page summary that lists key project information and also identifies issues that are already impacting the project, if any. The project manager or product owner may use this form at the beginning of a project to summarize information known to the team at the start of the project.

The first half of the Project or Issue Submission Form includes data fields to identify and remind team members and stakeholders of the business opportunity addressed by the solution under development, and the business goals, scope of work, and cost constraints of the project. Note that in-scope and out-of-scope items are specifically identifies to clearly define the boundaries for the project. The second half of the form provides fields to identify and describe issues, their impact, and recommendations for resolution.

Recall from previous discussions that the difference between a risk and an issue is that a risk is something that might happen and impact our project, whereas an issue is something that has already happened. In an ideal situation, the development team will already have anticipated the potential for the issue and developed one or more contingency plans to resolve the problems. Unfortunately, that is not always the case, and the team must react proactively to develop and execute a plan to address the problems as they occur.

Additional entries are provided to identify a description of the project, any critical issues, and alternatives and recommendations that may provide a viable path.

The data fields within the project or issue submission form include:

- General project information
- Opportunity statement
- Business goals
- Project scope
- Projected costs
- Project issue or problem identification information
- Description of issues
- Impact of issues/ problems
- Alternatives and recommendations

YOUR COMPANY LOGO

Project or Issue Submission Form
Project Name
Version

General Project Information (for new projects)	
Project Name:	
Submitted By:	
Contact Name and Tel # (if different):	
Date Submitted:	

Opportunity Statement (for new projects)

Describe in detail the business problem or opportunity addressed by the project or program.

- *Include the business drivers, key systems impacted, and the programs, projects, applications, and departments that are impacted.*
- *Identify the way the project fits with business and corporate strategic goals.*
- *Identify the risk(s) of not doing the project or program.*

Business Goal (for new projects)

List and describe the goals of the project, for example:

- *To reduce costs*
- *To increase efficiency.*
- *To satisfy regulatory requirements*
- *To comply with IT Governance mandates*
- *To satisfy internal/external auditors.*

Project Scope (for new projects)

Provide a high level of the processes/systems that will be impacted by the project, and then identify the in-scope and out-of-scope processes/systems.

Projected Costs

☐	$2,500,000-$5,000,000	☐	$250,000-$500,000
☐	$1,000,000-$2,500,000	☐	$100,000-$250,000
☐	$500,000-$1,000,000	☐	< $100,000

Project Issue or Problem Information

Issue Number	Submitter By (name)	Date Submitted
Issue Status	**Issue Assigned To**	**Est. Completion Date**

Description of Issue

Provide sufficient information to identify and track the issue.

Impact of Issue or Problem

How does this problem or issue impact the project? Describe any risks for not correcting the problem or issue.

Alternatives and Recommendations

Describe any alternatives or work-around recommendations.

PROJECT TEAM DEFINITION

This document identifies project stakeholders and participants, and provides their contact information, roles on the project, and their organizational affiliations. This may not seem to be a big deal, but the value of this form increases dramatically when multiple contractors and employees from different functional organizations are involved the support of a program or project.

The listing of stakeholders is just as important as listing the direct project participants. I define stakeholders as anyone whose opinions and actions can impact the program, both positively and negatively. Their opinions and input matter regardless of whether the team is using agile or traditional software systems development practices.

The project team consists of resources assigned to help develop the products/ deliverables and to support the work objectives of the project. The project team can consist of members who all come from the same organization or department. However, on large enterprise and highly integrated systems, the project is more likely supported by a cross-functional team with assigned members working temporarily, as either part-time or full-time resources, on loan from other functional departments and contract or partner organizations.

The Project team as a whole must:

- Have requisite skills to analyze the requirements and plan the work
- Have requisite skills to perform assigned work
- Accomplish their assigned work
- Stay within the approved constraints of time, budget, scope, and quality
- Communicate product and project status
- identify and mitigate risks and resolve issues

The Project Team Definition contains the following data fields:

- Stakeholders
  - Stakeholders' names
  - phone numbers
  - Business or technical area (e.g., functional areas)
  - Project role
- Project Participants
  - Team members' names
  - Phone numbers
  - Organizations/ work areas (e.g., functional areas, contractors or external partners)
  - Project role (and responsibilities)

Project Team Definition
Project Name
Version

YOUR COMPANY LOGO

Stakeholders

Stakeholder's Name	Phone #	Business / Technical Area	Role
			Business Sponsor / Project Champion
			Business Owner
			IT Sponsor
			Project Manager
			Etc.

Project Participants

Team Members Name	Phone #	Organization / Work Area	Role
			Project Manager
			IT Team Lead
			Business Analyst
			Systems Analyst
			Database Administrator
			Developer
			Tester
			Business Functional Manager
			Subject Matter Expert
			Etc.

STAKEHOLDER IDENTIFICATION LIST

Projects are often highly politicized endeavors. One of the reasons for this, from an internal perspective, is that there are seldom sufficient financial and human resources to do all the work that is identified to support the strategies and mission of the business. As a result, some projects won't get funded - which can cause hurt feelings and put targets on the backs of those managers whose projects were funded. Every action and missed performance objectives of those teams will be highly scrutinized.

Internal stakeholders include the managers and users of the project's deliverables. They need to be confident that the deliverables will meet their needs. External stakeholders, those outside the customer and end-user community, may view the project as positively or negatively impacting their lives or business interests. Regulatory agencies are another form of external stakeholders.

If we don't know who the stakeholders are, or their perceptions on how they might be impacted by our project, we may find them acting out against our team's efforts. Therefore, we not only need to know who our stakeholders are, we also need to know how our project potentially impacts them and what their motivations are.

Astute product owners and project managers understand that they must develop a stakeholder management plan. A stakeholder management plan includes the identification and analysis of potential stakeholders and documents relevant information, such as their interest, involvement, influence, and potential impact on the project's or product's success. Equally important, the stakeholder management plan defines actions to ensure stakeholders are aware of and participate in project team plans and decisions that might affect them.

The Stakeholder Identification List provides a tool to identify the people who could affect or be affected by our project, to analyze stakeholder's expectations and their impact on the project, and to develop specific strategies and tactics for effectively engaging stakeholders to reflect the stakeholder's interest and involvement in the project.

The Stakeholder Identification List provides the following data fields:

- **Stakeholder Identification**
  - Name,
  - Title,
  - Potential role,
  - Level of knowledge,
  - Level of commitment to the project
- **Power/ Interest Determination**
  - Name,
  - Organization,
  - Power level (1-10),
  - Interest level (1-10),
  - Impact (power level X interest level)

YOUR COMPANY LOGO

Stakeholder Identification List
Project Name
Version

3 Stakeholder Identification

Describe the individual and title, potential role that stakeholder will play, their level of knowledge of the issues, and the level of commitment each stakeholder is willing to engage to support the project.

To "fill in" the table, consider the following:

- *Potential Role – In what issues or activities will the individual participate?*
- *Level of Knowledge – How much does this individual understand the issues involved in the project?*
- *Level of Commitment – To what level will this individual support or oppose the project?*

Name	Title	Potential Role	Level of Knowledge	Level of Commitment
B. Jones	*Manager – Business Group*	*Provides business rules for calculating ROI.*	*Extremely knowledgeable in project – works directly below project sponsor*	*Strongly supports project but hesitant to use offshore resources for development, based on previous experiences.*

4 Power / Interest Determination

List the people, their organization, their power (from a scale of from 1 to 10 where 10 is the highest), their interest in the project (1-10), and their impact on the project (Power Level x interest Level).

Name	Organization	Power Level (1-10)	Interest Level (1-10)	Impact on Project (Power Level x Interest Level)
B. Jones	*Business Group*	*9*	*5*	*45*

ON-BOARDING CHECKLIST

This template provides a tool to document the process necessary to successfully onboard an employee or contractor. Areas that must be addressed include job description, financial/ salary, benefits, conditions of employment, functional department and management assignments, training, security clearances, drug testing and background checks, duties and responsibilities, federal regulations and tax requirements, and other legal requirements.

The On-Boarding Checklist includes following data fields:

- **Employee Information**
  - Employee name
  - Position name
  - Personal phone
  - Personal email address
  - Department
  - Manager/Supervisor
  - Work phone
  - Work email address
- **Orientation**
  - Introduction to team
  - Job assignment and training
  - Job description and expectations
- **Employment Contract**
  - Job title
  - Schedule
  - Duration of employment
  - Salary
  - Benefits package
  - Duties and responsibilities
  - Termination conditions
- **Legal Requirements**
  - W-4 form (W-9 for contractors)
  - I-9 Employment Verification Form
  - Direct deposit form
  - E-Verify System
  - Anti-Harassment documentation

YOUR COMPANY LOGO

On-Boarding Checklist
Project Name
Version

Employee Information			
Employee Name:			
Position Name:		**Position Number:**	
Work Phone:		**Email Address:**	
Department:			
Manager/ Supervisor			
Work Phone:		**Email Address:**	

	Orientation	Date Completed
☐	Introduction to team	
☐	Job assignment and training	
☐	Job description and expectations	

	Employment Contract	Date Completed
☐	Job title	
☐	Schedule	
☐	Duration of employment	
☐	Salary	
☐	Benefits package	
☐	Duties and responsibilities	
☐	Termination conditions	

	Legal Requirements	Date Completed
☐	W-4 form (W-9 for contractors)	
☐	I-9 Employment Verification Form	
☐	Direct Deposit Form	
☐	E-Verify System	
☐	Anti-Harassment documentation	

 Document: 1375 Page 1 of 4

www.SDLCforms.com

INITIATE PROJECT CHECKLIST

Project initiation processes vary from one organization to another and are highly dependent upon the individual requirements of customers, executive sponsors, the scope of the project, and organizational policies. As a result, the initiate project checklist is one that must be tailored to meet the needs of both the organization and the individual project.

The initiate project checklist provides various checkoff items to verify the completion of required project initiation tasks before the team starts to work on the objectives of the project. The checklist provides a tool to ensure nothing is forgotten and to confirm to executive sponsors and customers the completion of mandatory startup functions.

The Initiate Project Checklist contains the following topics:

- **Purpose**
  - State purpose of the review process and checklist
- **Systems Development Life Cycle**
  - Project Concept Phase Checklist
  - Project Definition
  - Business Case
  - Feasibility Study
  - Project Charter
  - Project Team Set up
  - Other?
- **Project Initiation/ Project Start Phase Checklist**
  - Project Definition and Requirements
  - Project Planning
  - Scope and Trade-offs
  - Scheduling
  - Costs
  - Estimating
  - Team definition
  - Communication
  - Other?

YOUR COMPANY LOGO

Initial Project Checklist
Project Name
Version

Note: Text displayed in blue italics is included to provide guidance to the author and should be deleted before publishing the document. In any table, select and delete any blue line text; then click Home→Styles and select "Table Text" to restore the cells to the default value.

1 Purpose

The purpose of the Initial Project Checklist is to provide sample information to use and to verify that major initial project functions and tasks have been completed within the Concept Phase.

2 Systems Development Lifecycle

The following table shows a quick summary of the systems development lifecycle (SDLC).

Phase #	Phase Name	Description
0	Concept	Project Definition Business Case Feasibility Study Project Charter Project Team Setup.
1	Initiation	Project Definition and Requirements Planning Scope and Tradeoffs Scheduling Costs Estimating Team Communication.
2	Development / Execution	Ongoing Planning Specifications Design Development Tracking and Control Reviews Testing.
3	Approval	User Acceptance Testing Implementation Planning.
4	Delivery	Project Implementation

 Document: 1400 Page 4 of 8

www.SDLCforms.com

PROJECT RESOURCE PLAN

This document provides a centralized source for definition of all resources required for a project, including project team size, required resources, facility needs, resource types and sources, project team organization, resource assumptions, risks and mitigations.

The Project Resource Plan contains the following data fields:

- **Project team size**
  - Required resources / skill sets
  - resource type
  - source of personnel
  - quantity
- **Facility resources**
  - facility type
  - need date
  - quantity
- **Personnel resource profile**
  - Resource type
  - Source
  - Rate / Hour
  - Hours by month
  - Total hours
  - Total Dollars
- **Project team organization**
- **Resource assumptions, risks and mitigations**
- **Stakeholder approvals.**

YOUR COMPANY LOGO

Project Resource Plan
Project Name
Version

5 Resources by Resource Type

This section of the document will identify the resource types, source, hourly rate, hours by month, total hours and total dollars for the project.

Resource Type	Source	Rate / Hour	Hours (by Month)												Total Hours	Total Dollars
			1	2	3	4	5	6	7	8	9	10	11	12		
UX Designer	*Internal*	*$60.00*	*175*	*175*											*350*	*$21,000*
Developer	*Internal*	*$65.00*			*175*	*175*	*175*	*175*							*700*	*$45,500*
Developer	*Consultant*	*$95.00*			*175*	*175*	*175*	*175*	*175*	*175*	*175*	*175*			*1400*	*$133,000*
Trainer	*Internal*	*$50.00*											*150*		*150*	*$7,500*
Technical Writer	*Internal*	*$50.00*											*150*		*150*	*$7,500*
														TOTALS:	*2750*	*$214,500*

 Document: 1450 Page 7 of 9

www.SDLCforms.com

This page is intentionally left blank

Project Planning Phase
Overview

The planning phase is the bulk of where the project and technical management planning work occurs. The project management planning work has a focus on organizing and scheduling work and developing project level processes to manage each of the following requirements:

- Agreements
- Change requests
- Contractors
- Decisions
- Internal and external quality reviews (e.g., SQA and IV&V reviews)
- Project budgets
- Project resources
- Project risks and issues
- Project procurements
- Reporting
- Suppliers

The technical management planning activities have a focus on implementing project-level processes that support development and operational needs. As an example, technology planning implements high-level strategies to support version control and configuration management, testing, architecture and design assessments, systems integration, software quality assurance (SQA), verification and validation requirements, and operational, maintenance and support requirements. However, since many of the technical processes map directly to the SDLC Phases, the technical planning artifacts (methods and tools) are managed within the appropriate SDLC phases, by type of work.

This does not have to be a particularly onerous task, though it takes some time to pull the information together, and review the proposed plans with customers, project team members, and other key stakeholders. The key is to have reusable boilerplate content within your SDLC process repository where you can pull the information on plans and activities that are most relevant to your project.

Forms & Templates

PROJECT MANAGEMENT OFFICE (PMO) CHECKLIST

The Project Management Office Checklist provides a tool to initially find and evaluate the appropriateness of corporate defined and approved conditions necessary to manage projects assigned to the PMO. The PMO Checklist conditions map to specific activities and tasks performed by members of both the PMO and their assigned project teams, the execution of which will satisfy the stated conditions within the checklist.

Larger and heavily regulated organizations can expect to have more predefined conditions than smaller ones. Still, not every activity listed is appropriate on every program, and the program manager, working with their PMO team and assigned project managers, must identify the activities and tasks most applicable for the scope of work assigned to the program. Once the initial assessments are completed, the PMO team will work with the project teams under their management to ensure all the conditions are satisfied.

When developing a PMO checklist it's important to bear in mind the functions of the office. Basic PMO functions include:

- Implementation of corporate approved project management standards and policies
- Project level training and coaching on project management practices
- Assignment of qualified project management professionals
- Implementation and oversight of lean-agile practices to maximize efficiencies, delivery speed, and customer responsiveness
- Assignment and oversight of human resources assigned to the project teams from functional departments and contractors
- Coordinating corporate resources for maximum efficiencies across the portfolio of assigned projects
- Providing operations and administrative support to assigned project teams

The PMO checklist contains the following data fields:

- Purpose and intended audience
- Organizational responsibilities:
- Project Management Office (PMO)
    - Project managers
    - PMO framework, interfaces and tools illustration
- PMO framework, interfaces and tools Checklist
- Data required by Project Managers
- Project manager's toolset illustration
- Project manager's interfaces and toolset checklist.

Project Management Office (PMO) Checklist
Project Name
Version

3.1 PMO Framework and Tools Checklist

Table 1-1 provides a checklist of the minimum interfaces and tools necessary for successful PMO execution. If these capabilities do not exist, the project managers will encounter difficulties satisfying their statusing and reporting requirements.

Check Item	Interface or Tool	Description
☐	PMO Overview	This is a set of instructions for defining and using the tools available to PMO, business units and project management personnel. Too often, this critical information dissemination is overlooked in many organizations.
☐	Funding Sources	Identifies the funding source and dollar value of all projects, and in some organizations will identify the types of funding, for example, capital vs. expense.
☐	Project Staffing/ Responsibilities	Identifies personnel assigned and their role in each project.
☐	Business Units/ Stakeholders	Identifies the business units, key personnel, and the stakeholders who will approve each project.
☐	PMM/SDLC Documents	This is the well documented Program Management Methodology (PMM) employed by the company including templates and documents utilized in the System Development Life Cycle (SDLC) available for downloading by the business analysts and project managers
☐	Work Process Flow/ Approvals	Provides a mechanism for process work flow and approval for each document in the life cycle (example: with the automated use of SharePoint), with appropriate alerts and notifications when approvals are delinquent.
☐	Enterprise Architecture	Project designs must be reviewed by the enterprise architecture organization to ensure compatibility with long-term strategic and architecture goals, and as a repository for all standards (such as screen design or coding standards).
☐	Project Issues	A list of project issues, with associated status, can be assessed by the project managers and business units and updated by the project managers.
☐	Project Risks	Project risks can be identified and addressed by the project managers and business units and updated by the project managers.
☐	Project Dependencies/Interfaces	Separate lists of project dependencies and interfaces are necessary for collaboration and actions between project managers, ensuring that these items are readily traceable.
☐	Executive Management Reporting	To ensure project success, frequent and concise project status updates to executive and senior IT management, from data provided by individual project managers, is critical. This capability must include the ability to generate flexible and customizable reports to satisfy senior management and PMO on-demand requests.

www.SDLCforms.com

PROJECT MANAGEMENT PLAN TEMPLATE

The Project Management Plan template describes the project management processes, methods and tools employed within a project; which are often incorporated as a set of project management subplans, or chapters within the PM Plan. Projects have the following characteristics:

- A specific and relatively unique scope of work
- A temporary duration with an explicit beginning and an end
- Have specific deliverable requirements in the form of products, services, or outcomes
- Have incomplete information that must be defined over the course of the project's duration
- Often requires cross organizational resources to execute.

The project management plan is as detailed as it needs to be, but not one word more, or it won't be read or followed. However, regardless of the scale of the project, project processes span the following activities:

- Project initiation activities
- Scope and schedule the work
- Monitor progress against the schedule plans
- Manage risks and issues
- Deliver the right set of products and services to specifications
- Close out the project-based work and reassign resources.

It is good practice to have a formally agreed and version-controlled project management plan approved in the early stages of the project and applied and updated throughout the project. When completed, the project management plan should be reviewed and approved by the project's clients, executive sponsors, the assigned project manager, and other key stakeholders.

The project management plan typically includes the following chapters or sub plans:

- Scope Management Plan
- Schedule Management Plan
- Financial Management Plan
- Quality Assurance Management Plan
- Resource Management Plan
    - To include personnel, equipment, tools, etc.
- Communications Management Plan
- Project Change Management Plan
- Risk Management Plan
- Procurement Management Plan
- Other relevant planning information

YOUR COMPANY LOGO

Project Management Plan Template
Project Name
Version

Note: Text displayed in blue italics is included to provide guidance to the author and should be deleted before publishing the document. In any table, select and delete any blue line text; then click Home→Styles and select "Table Text" to restore the cells to the default value.

1 Overview

This Project Management Plan Template is a complete aggregation of project information and sub-plans for the entire project, which can include the following:

- Project Overview
- Project Scope
- Project Approach
- Project Estimation
- Project Milestone Review
- Project Resource Plan
- Project Data Management Plan
- Risk (Assessment & Mitigation) Plan
- Project Measurements
- Project Communications Plan
- Monitoring and Control Plan/ Documentation Plan
- Supplier Agreement Plan
- Configuration Management Plan – Scope Statement
- Process and Product Quality Assurance Plan
- Finance Plan - Scope Statement.

1.1 Audience

This document is intended for use by all associates and stakeholders involved in the project.

What is your reader's technical level, experience level, interest in background details?

STATEMENT OF WORK (SOW)

SOW provides guidance from customers or executive sponsors on the scope of work approved for the project and expected deliverables. A SOW may also provide information on key contacts, business drivers and objectives, quality and technical standards, identified risks, and delivery milestones and budgets.

In a contract setting, the SOW is often incorporated as an appendix within the initial request for quote (RFQ) and the final contract agreements. In this manner, the scope of work and project constraints specified within the SOW becomes part of the agreement.

The SOW must contain sufficient detail, so all stakeholders understand the required scope of work, the duration of work, and identification of the specific deliverables. This is just as true when it's part of our RFQ as it is for a final contract agreement as the SOW is the basis of information and upon which contractors and suppliers build their proposals.

The Statement of Work contains the following data fields:

- **Introduction**
  - Stakeholders and Contacts
- **Business Objectives**
  - Business Need or Opportunity
  - Business (Product) Description and Solutions
  - Deliverables
  - Software Verification and Validation
- **Project Description**
  - Completion Criteria
  - Risk Assessment
  - Constraints
  - Dependency Linkages
  - Impact
  - Measures of Project Success
  - Roles and Key Project Participants
- **Project Estimates**
  - Estimated Schedule
  - Resource Requirements - Team and Support Resources
  - Estimated Costs
- **Project Controls**
  - Risk / Contingency Management
  - Issue Management
  - Change Management

YOUR COMPANY LOGO

Statement of Work
Project Name
Version

Note: Text displayed in blue italics is included to provide guidance to the author and should be deleted before publishing the document. In any table, select and delete any blue line text; then click Home→Styles and select "Table Text" to restore the cells to the default value.

1 Introduction

Provide a brief summary of the project using the Project Charter and Business Case as a basis for this text if available.

1.1 Referenced Documentation

Provide information about all documentation referenced in this document, e.g., number, title, version, and date.

Document Number	Version	Date	Document Name

1.2 Stakeholders and Contacts

Provide information about the stakeholders (e.g., sponsors, project and functional managers) and contacts for the project.

Stakeholders

Name	Company or Department	Title	Phone	Email address

Contacts

Name	Company or Department	Specialty, Area, or Role	Phone	Email address

PROJECT APPROVAL DOCUMENT

The Project Approval Document formalizes the approval for any project related deliverable. The project approval document can be used to approve the start of work on a new deliverable item, and it can be used to demonstrate the customer has approved the final deliverable after the work is completed.

The project approval document is signed by the executive sponsor or customer responsible for the deliverable item. However, other project team members and stakeholders may initially signoff to indicate they have reviewed and approved the deliverable request or successful completion of the deliverables.

The Project Approval Document contains the following data fields:

- Overview
    - Topic
    - Description
- Project Description
- Approval Information
    - Responsibility/ Organization
    - Name
    - Approved Signature
    - Date

Project Approval Document
Project Name
Version

Note: Text displayed in blue italics is included to provide guidance to the author and should be deleted before publishing the document. In any table, select and delete any blue line text; then click Home→Styles and select "Table Text" to restore the cells to the default value.

This form formalizes approval for the <Name of Project> Project.

Topic	Description
Project Name	
Project Number	
Department or Business Unit Name	
Department or Business Unit Cost Center	
Sponsor Name and Telephone Number	
Project Manager Name and Telephone Number	

Project Description
Provide a description of the project.

Provide approval information in the following table.

Responsibility / Organization	Name	Approved Signature	Date	Comments (e.g., major or minor, cost, not approved)
Project Sponsor				
Business Owner				
Project Manager				

Confidential – ©2015 Documentation Consultants (www.SDLCforms.com) Document: 2100 Page 1 of 2

www.SDLCforms.com

COST ESTIMATING WORKSHEET

This Excel spreadsheet provides the opportunity to estimate and budget various IT costs. A common planning heuristic is to analyze the requirements to a degree sufficient to have confidence that the estimate is within a range of -10% - +25% of actual costs. The reason the range is higher on the positive side is that, from a risk perspective, if you are the supplier, you want to make sure you have not underestimated the scope of work.

If you are an extremely competitive situation, and you're afraid the higher-end costs will put you out of the bed, then take the time to ask more questions and get a better handle on the requirements so that you can provide a more accurate estimate. If you believe your costs are fair but still high, be prepared to defend your position with data to back up and justify your cost estimations. It does not do your company, nor your customers, a service to underbid the costs on a project and put your company at risk.

The Project Cost Estimate breaks item costs out into the following high-level cost items:

- IT Resources
  - IT Resources
  - External professional services
  - Hardware/communications and maintenance
  - Software license fees and maintenance
  - Other fees (travel, training, etc.)
- FORMULAS

The embedded formulas include the following:

- IT Resources Project Cost = Staff Hours X Hourly Rate
- Professional Services Cost = Hours X Hourly Rate
- Hardware / Communications & Maintenance Costs Subtotal = Total of individual items
- Software Licenses & Maintenance Costs Subtotal = Total of individual items
- Other (Travel, Supplies, etc.) Costs Subtotal = Total of individual items
- Total = all of the above.
- Risk / Contingency Reserve = 15% (user definable) of Total
- Grand Total = All of the above.

Cost Estimating Worksheet
Project Name
Version

Budget Level Estimate: -10% to +25%

Cost Items	Project Cost			Yearly Maint/Admin/Support Cost		
IT Resources ***(Resource Class Totals from Project Labor Estimates Sheet)***	Staff Hours	Hourly Rate	Staff Cost	Staff Hours	Hourly Rate	Staff Cost
			$0			$0
			$0			$0
			$0			$0
Subtotal			**$0**			**$0**
External Professional Services	Hours	Hourly Rate	Prof Svc Cost	Hours	Hourly Rate	Prof Svc Cost
			$0			$0
			$0			$0
			$0			$0
Subtotal			**$0**			**$0**
Hardware / Communications & Maintenance			Cost			Cost
Subtotal			**$0**			**$0**
Software Licenses & Maintenance			Cost			Cost
Subtotal			**$0**			**$0**
Other (Travel, Supplies, etc).			Cost			Cost
Subtotal			**$0**			**$0**
TOTAL			**$0**			**$0**
Risk / Contingency Reserve %		15.0%				
GRAND TOTAL			**$0**			**$0**

www.SDLCforms.com

DEVELOPMENT ESTIMATING WORKSHEET

This Excel spreadsheet provides a tool to estimate development costs for prototyping, user interfaces, reports, databases, tables, objects, and integration / jobs. The Development Estimating Worksheet consists of two tabs: Development Estimation Worksheet and a Servers Pricing Calculator.

This Development Estimating Worksheet Excel spreadsheet consists of the following sections:

- **Development Estimation Worksheet**
- **Recommended Staffing,**
- **Software & Licensing costs**
- **Ongoing Support Costs**
- **Summary Total Costs.**

The Development Estimation Worksheet provides formulas to calculate the following data fields:

- **Estimating Topics**
  - Prototypes
  - User Interface / Reports
  - Tables
  - Objects
  - Jobs
  - Conversion / Migration
  - Configuration / Setup
  - Documentation & Discovery
  - Miscellaneous Tasks...
- **Calculate Total Development Resources**
  - Recommended Staffing
- **Recommended Staffing**
  - Software and Licensing Costs
- **Software and Licensing Costs**
  - Ongoing Support Costs
  - Ongoing Support Costs
  - Summary Total Costs
- **Servers Pricing Calculator**
  - Windows 2000/2003/2010 Servers
  - Microsoft Active Directory Servers
  - Microsoft IIS Servers
  - Microsoft Exchange Servers
  - Microsoft SQL Servers
  - Oracle Database Servers.

This form to be modified to add any other cost items within your proposed solution offering.

Development Estimating Worksheet
Project Name
Version

Servers Pricing Calculator

Updated As Of: mm/dd/yyyy

Customer Discount Rate:	0%
Total Cost for All Licensing:	$ -

Windows 2000 \ 2003 Servers (Base server monitoring only)

Total Number of Servers:	0		
Required Monitoring Components	Cost Per License	Discounted Price	Sub-Total Cost Per Component
Base Monitoring Agent:	$ -	$ -	$ -
Yearly Maintenance:	$ -	$ -	$ -
Performance Profiler for Windows:	$ -	$ -	$ -
Performance Profiler Yearly Maintenance:	$ -	$ -	$ -
Analysis Center License:	$ -	$ -	$ -
Analysis Center Yearly Maintenance:	$ -	$ -	$ -
		Total Cost for all Servers:	$ -

Microsoft Active Directory Servers

Total Number of Servers:	0		
Required Monitoring Components	Cost Per License	Discounted Price	Sub-Total Cost Per Component
Base Monitoring Agent:	$ -	$ -	$ -
Yearly Maintenance:	$ -	$ -	$ -
Monitoring:	$ -	$ -	$ -
Yearly Maintenance:	$ -	$ -	$ -
Performance Profiler for Windows:	$ -	$ -	$ -
Performance Profiler Yearly Maintenance:	$ -	$ -	$ -
Analysis Center License:	$ -	$ -	$ -
Analysis Center Yearly Maintenance:	$ -	$ -	$ -
		Total Cost for all Servers:	$ -

Microsoft IIS Servers

Total Number of Servers:	0		
Required Monitoring Components	Cost Per License	Discounted Price	Sub-Total Cost Per Component
Base Monitoring Agent:	$ -	$ -	$ -
Yearly Maintenance:	$ -	$ -	$ -
Monitoring:	$ -	$ -	$ -
Yearly Maintenance:	$ -	$ -	$ -
Performance Profiler for Windows:	$ -	$ -	$ -
Performance Profiler Yearly Maintenance:	$ -	$ -	$ -
Analysis Center License:	$ -	$ -	$ -
Analysis Center Yearly Maintenance:	$ -	$ -	$ -
		Total Cost for all Servers:	$ -

Microsoft Exchange Servers

Total Number of Servers:	0		
Required Monitoring Components	Cost Per License	Discounted Price	Sub-Total Cost Per Component
Base Monitoring Agent:	$ -	$ -	$ -
Yearly Maintenance:	$ -	$ -	$ -
Monitoring:	$ -	$ -	$ -
Yearly Maintenance:	$ -	$ -	$ -
Performance Profiler for Windows:	$ -	$ -	$ -
Performance Profiler Yearly Maintenance:	$ -	$ -	$ -
Analysis Center License:	$ -	$ -	$ -
Analysis Center Yearly Maintenance:	$ -	$ -	$ -
		Total Cost for all Servers:	$ -

Microsoft SQL Servers

Page 1 of 2

www.SDLCforms.com

PROJECT CAPITAL VS. EXPENSE COSTS WORKSHEET

This worksheet provides a tool to estimate various capital and expense costs within a project including IT resources, external professional services, hardware, communications, software licenses, and travel and supplies.

The Project Capital and Expenses Costs can then be tracked against the budget allocated for the project.

The Project Capital Vs. Expense Costs contains the following data fields:

- **Capital/Expense Worksheet**
  - Project Management
  - Pre-Development Costs
  - Development Costs
  - Post-Development Costs
  - Hardware/Communications and Maintenance
  - Software Licenses
  - Software Maintenance
  - Other (Travel, supplies, etc.)
  - Subtotal
  - Contingency
  - Grand Total.

Capital / Expense Worksheet

Provides an itemization of estimated capitalized and expensed costs for implementing the proposed solution.

CAPITAL / EXPENSE WORKSHEET				
Functional Role	**Work Order #**	**Account #**		
Project Management				**Expense**
Project Manager - ALL PHASES				
Subtotal				$0
Pre-Development Costs				**Expense**
Developer Requirements Definition				
System Analyst (SA) Requirements Definition				
Quality Assurance (QA) Requirements Definition				
Business Analyst (BA) Requirements Definition				
Subtotal				$0
Development Costs			**Capital**	
Developer Design, Develop, Test, Implement				
SA Design, Develop, Test, Implement				
Tech Writer Design, Develop, Test, Implement				
QA Design, Develop, Test, Implement				
BA Design, Develop, Test, Implement				
Subtotal			$0	
Post-Development Costs				**Expense**
Support				
Training				
Subtotal				$0
Hardware / Communications & Maintenance			**Capital**	
Subtotal			$0	
Software Licenses			**Capital**	
Subtotal			$0	
Software Maintenance			**Capital**	
Subtotal				$0
Other (Travel, Supplies, etc)			**Capital**	**Expense**
Travel During Development				
Other Travel				
Subtotal				$0
			Capital	**Expense**
SUBTOTAL			$0	$0
Contingency (15%)			$0	$0
GRAND TOTALS			$0	$0

www.SDLCforms.com

CONFIGURATION MANAGEMENT PLAN

The Configuration Management (CM) Plan informs project team members and stakeholders on the corporate CM policies and processes governing the project. CM practices ensure the organization properly versions, tracks, archives, and maintains over the life of a product all configuration items and other artifacts related to each release of the product. Organizations that do not employ rigorous CM practices can be assured that long-term sustainability of their products is compromised. That means not only will sustainability costs increase; it will also be increasingly difficult if not impossible to enhance and maintain their legacy products.

The CM plan is used to ensure proper CM practices are followed, identify the CM tools used by the project team, and how configuration items will be identified and managed across the project and the lifecycle the deliverable products.

The plan contains the following data fields and tables:

- **Organization**
  - Identify Individuals with CM responsibilities
- **Activities and Responsibilities**
  - Setting up project activities
  - Project planning life cycle activities
  - Management activities
- **Decision process and responsibilities**
  - Project activities
  - Process audit activities
- **Configuration Control Board (CCB) responsibilities**
- **Individual Configuration Control Board responsibilities**

Configuration Management Plan
Project Name
Version

4 ORGANIZATION

Describe the Configuration Management organization. Software configuration is managed by team members. Responsibilities are shared between:

- *Software Configuration Manager (SCM)*
- *Project Manager (PM)*
- *Quality Manager (QM)*
- *Development Manager (DM).*

4.1 Activities and Responsibilities

List the functions required to manage the configuration of the software and the associated responsibilities.

Setting Up Project Activities	Individual Responsible
Identify configuration items	*Software Configuration Manager*
Install bug repository tool and set up a database	*Software Configuration Manager*
Define the configuration process	*Software Configuration Manager*
Project Planning Life Cycle Activities	**Individual Responsible**
Create version, write version delivery document	*Software Configuration Manager*
Approve reference configuration	*Software Configuration Manager*
Do configuration audits	*Software Configuration Manager*
Archive reference version	*Software Configuration Manager*
Do configuration audits	*Quality Manager*
Verify version to be delivered and authorize deliveries	*Project Manager*
Management Activities	**Individual Responsible**
Manage versions and archives	*Software Configuration Manager*
Manage configuration records	*Software Configuration Manager*
Manage reference space and its access control list	*Software Configuration Manager*
Manage quality reports	*Quality Manager*

 Document: 2300 Page 6 of 11

RISK INFORMATION DATA COLLECTION FORM

During the course of a project, potential risks can be identified by a myriad of sources. The Project Risk Information Data Collection Form's purpose is to provide a vehicle for capturing detail information on any of those risks for analysis and evaluation. Summary information from this data collection is then encapsulated in the Project Risk Analysis Plan for weekly management review.

The form contains the following topics:

- Section 1: Risk Identification
- Section 2: Risk Root Cause Analysis
- Section 3: Risk Evaluation
- Section 4: Risk consideration and response

Risk Information Data Collection Form
Project Name
Version

During the course of a project, potential risks can be identified by a myriad of sources. The Project Risk Information Data Collection Form's purpose is to provide a vehicle for capturing detail information on any of those risks for analysis and evaluation. Summary information from this data collection is then encapsulated in the Project Risk Analysis Plan for weekly management review.

Note: In any table, select and delete any blue line text; then click Home→Styles and select "Table Text" to restore the cells to the default value.

Risk Information Data Collection Form

Section 1: Risk Identification

Risk Source Name	Phone Number	Email Address

Risk ID	Date Risk Identified	Risk Owner	Risk Action Owner
RI-999	*99/99/9999*		

Risk Short Title	Risk Event Description

Section 2: Risk Root Cause Analysis

Risk Assessment	*If the Risk…*
Risk Impact	*Then the Risk Impacts…*

Section 3: Impacted Projects

Risk Affects	☐	*Project Name 1*	☐	*Project Name 2*	☐	*Project Name 3*	☐	*Program Level*
Impact Areas	☐	Cost	☐	Schedule	☐	Resource	☐	Scope
	☐	Quality	☐	Safety	☐	Reliability	☐	

RISK ANALYSIS TEMPLATE

The Risk Analysis template provides a tool to record identified risks and their related risk assessments over the course of the project. Specifically, the project's risk analysis template will help the team identify, record, evaluate, and manage risks and ensure the team has developed risk mitigation strategies and contingency plans for each identified risk. Risk mitigation strategies are what the team plans to do to avoid or minimize the impact of the risk, while a contingency plan is what the team plans to do if the risk in fact occurs despite the best efforts.

As part of the risk analysis method, the team should consider the following factors:

- Assess the impact of the risk, e.g., lack of resources, operational disruptions, access denial, procedural failures, quality problems; tasks take too long, cost over-runs, budget problems, and technical failure.
- Estimate the risk and probability of the event occurring.
- Manage the risk and crisis prevention.

The Project Risk Analysis template contains the following data fields:

- Purpose for Project
- Project Information
  - Project Description
  - Goals, Objectives, and Scope
  - Assumptions, Dependencies, and Constraints
- Risk Analysis Table - in which the risk threat is identified, the priority score established, the risk mitigation strategy and date are recorded, and the contingencies / triggers defined.

Project Risk Analysis
Project Name
Version

3 Risk Analysis Table

Use the table to keep track of potential risks that may jeopardize the project's success or completion date. In the initial columns, identify potential project risks.

- *Assign a percentage value to the Risk (this is the probability that a particular risk will occur).*
- *Fill in a number from 1 to 10 to indicate the Impact that an event will have on the project (10 has greatest impact), or copy the values from the Project Risk Information Data Collection Form.*
- *Calculate the Score by multiplying the Risk by the Impact.*
- *Identify the Risk Mitigation in the Strategy column (steps to take to alleviate potential risks and associated dates).*
- *Fill in the final column with a backup plan in case the risk strategy is insufficient to prevent a potential risk.*

#	Risk/Threat	Priority			Risk Mitigation Strategy		Contingency / Triggers
		Risk Probability (1-5)	Impact (1-10)	Score (Risk X Impact)	Strategy	Date	
R-1	*Insufficient Project Resources - staff*	*2.5*	*7*	*17.5*	*Assign a high priority to the project and notify dept. heads that some staff may need to be pulled from other projects.*	*Project Start Date*	*Contract for additional resources. Trigger (e.g., project scope enlarged)*

 Document: 2400 Page 6 of 6

www.SDLCforms.com

PROCUREMENT PLAN

The Procurement Plan provides procedures and information on how to acquire hardware, software, vendors or other needed items over the course of the project. The plan, as procurement needs are identified, documents what items must be acquired, when, and how. Moreover, the template must be modified by the organization to conform with procurement policies and governing regulations.

The Procurement Plan is used to provide information about the purchase of commercially available goods and services. It also provides decision support information governing how vendors are selected, managed, obtain funding approvals, build versus buy strategies, and who must be involved at each stage of the procurement process.

The Procurement Plan contains the following data fields:

- Introduction
  - Purpose, Scope, and Objectives
  - Project Background
  - Referenced Documentation
- Procurement Information
  - Customer / Contract Manager
  - Item(s) to be acquired and estimated value
  - Duration of contract or service
  - Market capability
  - Risks
  - Procurement timeline
- Roles and responsibilities
- Procurement Strategy
  - Pricing strategy
  - Dollar limits
  - Purchase Order requisition thresholds
  - Procurement method
  - Competitive solicitation

Procurement Plan
Project Name
Version

Document Number	Version	Date	Document Name

2 Procurement Information

Provide general procurement information in the following table.

Topic	Description
Customer / Contract Manager	*Contact information for the individual and organization requesting procurement.*
Item(s) to be Acquired and Estimated Value	*An estimate of the items and contract / service value over the life of the contract to cover the items or the contract duration in a service contract.*
Duration of Contract or Service	*The contract period from contract commencement to the delivery of the final output, including any options to extend the contract beyond the core period.*
Market Capability	*An evaluation of the competitive environment including suppliers with the capability and capacity to meet your requirements.*
Risks	*Identify the relevant risks involved, e.g.,* • *Business Risk, e.g., supplier's failure to deliver.* • *Service / Operational Risk, e.g., managed by a 3rd party to design, build, finance or operate.* • *External Risk, e.g., government regulations.*
Procurement Timeline	*When are the items needed?* • *Prepare a timeline for the activities if necessary, e.g., main tasks and milestone dates for the completion of such tasks.*
Roles and Responsibilities	*Who is responsible for planning and acquisition?* • *Evaluation Team including their strengths and weaknesses.* • *Prepare a matrix of roles and responsibility if*

 Document: 2450 Page 5 of 9

QUALITY PLAN

Managing project quality requires an approved quality plan encompassing three major quality processes, as identified below.

Quality Assurance – includes activities focused on the processes being employed to manage and deliver the solution to evaluate overall project performance on a regular basis. Quality assurance is a method to ensure the project will satisfy the quality standards and will define and record quality reviews, test performance and customer acceptance.

Quality Control – includes activities to verify that project management and project deliverables are of high quality and meet quality standards. Quality assurance also helps uncover causes of unsatisfactory results and establish lessons learned to avoid similar issues in this and other projects.

Quality Acceptance Criteria - identifies the major project deliverables, their acceptance criteria, and the responsible organizations and individuals who will confirm conformance of the deliverables with requirements and acceptance criteria.

The quality plan is developed and approved during the project planning phase to confirm major deliverables/milestones acceptance criteria and to manage approved project processes.

Quality Plan
Project Name
Version

Note: Text displayed in blue italics is included to provide guidance to the author and should be deleted before publishing the document. In any table, select and delete any blue line text; then click Home→Styles and select "Table Text" to restore the cells to the default value.

1 Purpose

Identify the project quality plan's purpose and how it meets specific project needs.

2 Quality Management Method

Use the following sections to identify the project quality plan components or modify these sections to meet your needs.

Managing project quality requires an approved quality plan encompassing three major quality processes identified below. The quality plan is developed and approved during the project planning phase to confirm major deliverables/milestones acceptance criteria and to manage approved project processes.

2.1 Quality Plan Processes

2.1.1 Quality Assurance

Quality assurance activities focus on the processes being employed to manage and deliver the solution to evaluate overall project performance on a regular basis. Quality assurance is a method to ensure the project will satisfy the quality standards and will define and record quality reviews, test performance and customer acceptance.

2.1.2 Quality Control

Quality control activities are performed continually to verify that project management and project deliverables are of high quality to meet quality standards. Quality assurance also helps uncover causes of unsatisfactory results and establish lessons learned to avoid similar issues in this and other projects.

2.1.3 Quality Control

Project team members and key stakeholders agree at the project planning stage on formal project processes and major deliverable acceptance criteria that will be used to evaluate final deliverable results before the results are formally approved.

PROJECT ORGANIZATION CHART

Larger projects and programs often involve participation of numerous people across the life cycle of the engagement. Project and program participants can include executive sponsors, customers, managers, team members, users, product owners, process owners, and other stakeholders who are potentially impacted by the programs and projects. It's important that the assigned project manager maintains and distributes a record of all potential project participants and their respective roles and responsibilities, as well as their contact information.

Typically, the roles and responsibilities of program and project participants are recorded in one or both of the following formats:

- **Project Contact Log/ List - that is usually implemented in a worksheet or document table format**
- **Project Organization Chart - that graphically portrays the working relationships across all assigned participants**

The Project Organization Chart, supports the second approach, and is a tool and graphical diagram that identifies all PMO and project personnel, sponsors, stakeholders, and business analysts, spanning the collaborating functional organizations such as Infrastructure, design, quality assurance, etc. However, don't forget to also include team members from participating contractors and other partnering organizations.

Most large companies and government agencies develop organizational charts to show, departments, individuals, and functional relationships within the enterprise. The Project Organization Chart is no different except the chart depicts the functional relationships at the program or project level.

The project organizational chart should include all pertinent personnel directly involved within the program or project, including (but not limited to):

- PMO
- Project managers
- Sponsors
- Business analysts
- Infrastructure
- Database design / reporting
- Quality Assurance
- Technical Support
- Any other pertinent organizations.

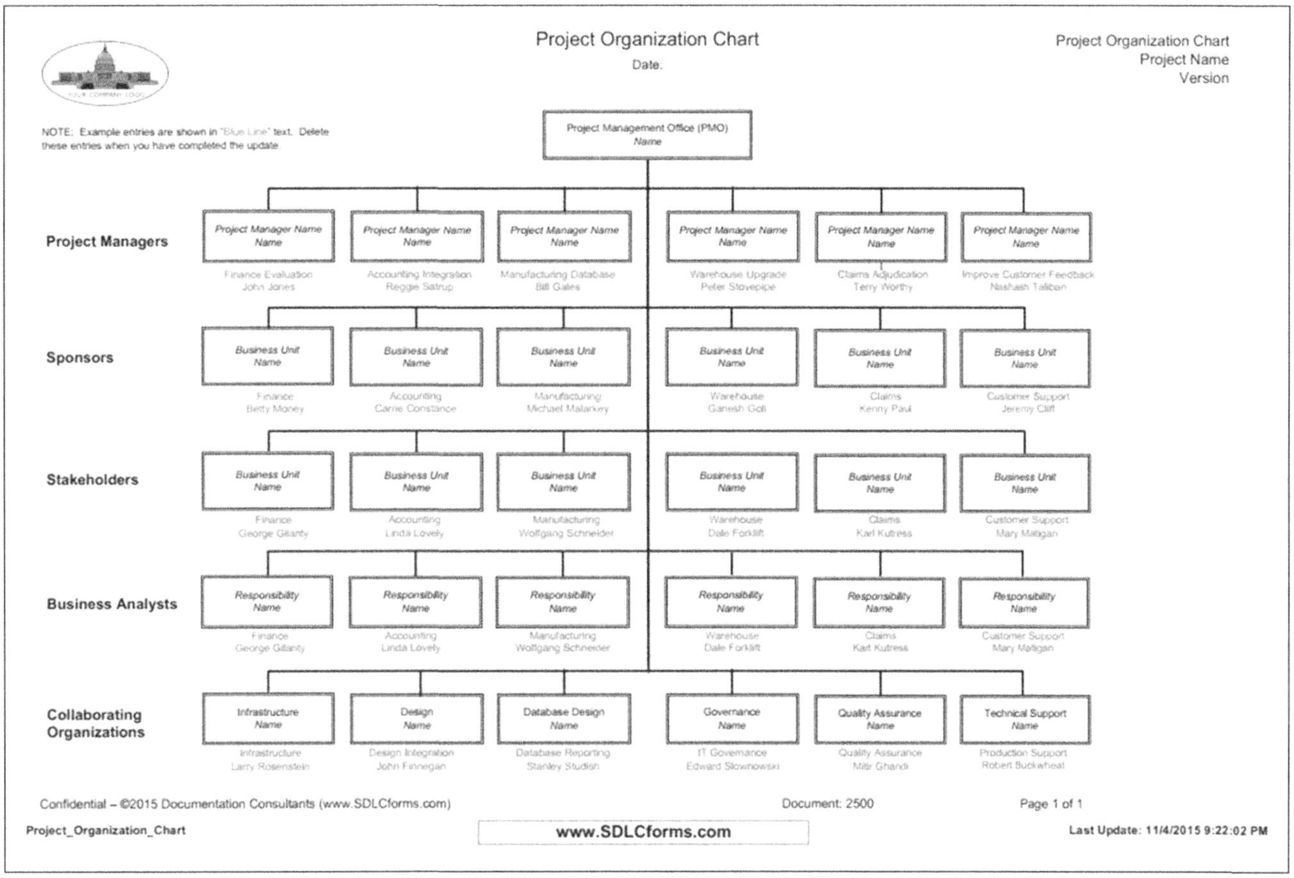
Project Organization Chart
Date:
Project Organization Chart
Project Name
Version
Project Management Office (PMO)
Name
Project Managers
Sponsors
Stakeholders
Business Analysts
Collaborating Organizations
Confidential – ©2015 Documentation Consultants (www.SDLCforms.com)
Document: 2500
Page 1 of 1
Project_Organization_Chart
www.SDLCforms.com
Last Update: 11/4/2015 9:22:02 PM

ROLES AND RESPONSIBILITIES MATRIX

The roles and responsibilities matrix serve a different function in the previously noted project organization charts and contact list. This tool is used to initially identify the types of roles and responsibilities necessary to support a program or project, and then identifies the functional departments and contractors that will provide the human resources to fill those positions.

A roles and responsibilities matrix list key project activities and details the responsibilities for each individual or role by functional department and/ or contractors. The matrix can optionally be created in a table or spreadsheet form within the document.

Projects come in different sizes and there are different ways and requirements on how people are organized. Multiple matrices are provided to help fill that need.

- Small projects: Little organizational structure is generally needed.
- Large projects: Many people are involved. It is important that people understand what they are expected to do, what role they are expected to fill, and the approval process.
- Responsibility Matrix: This technique is used to define the general responsibilities for each project role (i.e., who does what), communicates the roles to the team, and ensures people know what is expected from them.

The Roles and Responsibilities Matrix contains the following information and data fields:

- Purpose
- Setting Up a Responsibility Matrix
- Sample Matrix Roles and Responsibilities Description
- Roles and Responsibility Matrix
- Standard Roles and Responsibility Matrix
- RACI Roles and Responsibility Matrix

Note: RACI stands for the following work roles:

- Responsible - does the work
- Accountable (Approver) – The leader assigned to ensure completion of the work
- Consulted – subject matter experts who provide expertise and managers who need to review and approve work or changes in the scope of work
- Informed - usually executive sponsors, customers, and other stakeholders who need to stay abreast of the progress of the project, or any risks or issues associated with the project

Another popular version of a roles and responsibilities matrix is the RASCI Responsibility manager. The primary difference is the RASCI version includes a Support (S) role, which is assigned to anyone who is called in to help perform the work.

YOUR COMPANY LOGO

Roles and Responsibilities Matrix
Project Name
Version

2.1 Sample Matrix Roles and Responsibilities Descriptions

Roles	Description
Analyst	Primary liaison between technical areas, business partners or other project groups. The Analyst gains an understanding for current and future processes, documents business client requirements prior to developing and implementing solutions, and identifies costs, benefits, risks, and constraints. Note: This individual is often called a Business Analyst, Business Systems Analyst, Systems Analyst or a Requirements Analyst.
Change Control Group	The Change Control Group is generally composed of individuals authorized to approve changes to the project requirements, budget, and schedules.
Client	Project or service customers (individuals or groups).
Client Project Manager	Large projects can also have a primary business contact on the same level as the project manager representing the Information Technology organization.
Designer	The Designer reviews the business requirements and architecture and designs a solution to meet the business needs. Designer results are passed to development staff to build the solution based on the design specifications.
Functional Manager	Ensures project tasks are completed with resources under the manager's control within scope, time, budget, and quality constraints.
Governance Board	Executive Team in the organization that reviews projects to ensure they are meeting enterprise goals.
Performing Organization	Generally the sponsoring project organization whose personnel are most directly involved in doing the project work.
Program Manager	Individual in charge of management of a program (group of related projects) to achieve its objectives and benefits.
Project Management Office	A group or entity with responsibilities to manage those programs and projects under their responsibility.
Project Manager	The individual assigned to complete the project's objectives by coordinating and integrating activities across multiple functional lines, and managing stakeholder communications. The project manager also manages the project scope, time, cost, and quality.
Project Stakeholders	Individuals or groups that are actively interested (or have a stake) in the outcome of the project through its execution or implementation. They could be internal clients, management, employees or external stakeholders.

REQUIRED APPROVALS MATRIX

In the traditional SDLC model, there are many "gates" or "milestone reviews" that the program and project teams must get through in order to conclude a stage of work or proceed with the next stage of work. The required approvals matrix tracks the required approvals.

The Required Approvals Matrix provides a matrix of key project activities (e.g., functions, tasks, documents or phases), and who is responsible for approving them.

The Required Approvals Matrix contains the following data fields:

- **Purpose of the Project**
- **Sample Roles and Responsibilities**
- **Task Descriptions**
- **Approval Matrix**

Business Case Study	Project Plan	Operations Guide
Feasibility Study	Training Plan	Technical Test Plan
Cost / Benefit Analysis	System Design Document	User Acceptance Test Plan
Project Approval Document	Technical Design Document	Product Acceptance Document
Project Charter	Process Guide	Production Turnover Approval
Functional Requirements	Installation Guide	Project Feedback Analysis
Technical Requirements	Software User Guide	Modification Request
Requirements Traceability Matrix	System Administrators Guide	

Figure 5. Sample list of Approval Items

Required Approvals Matrix
Project Name
Version

2 Approval Matrix

The following table provides a matrix of functions or documents and those responsible to approve them.

Sample Matrix: Revise the table with the appropriate functions, tasks, documents, and phases.

Function or Document	Role or Name	Approval Name	Date Completed	Date Approved	Comments
Business Case Study	*Sponsor Managers*				
Feasibility Study	*Sponsor Managers*				
Cost / Benefit Analysis	*Sponsor Managers*				
Project Approval Document	*Sponsor Managers*				
Project Charter	*Sponsor Managers*				
Functional Requirements	*Sponsor Managers*				
Technical Requirements	*Proj. Mgr.*				
Requirements Traceability Matrix	*Proj. Mgr.*				
Project Plan	*Sponsor Proj. Mgr. Managers*				
Training Plan	*Proj. Mgr.*				
System Design Document	*Proj. Mgr.*				
Technical Design Document	*Proj. Mgr.*				
Process Guide	*Proj. Mgr.*				
Installation Guide	*Proj. Mgr.*				
Software User Guide	*Proj. Mgr.*				
System Administrators Guide	*Proj. Mgr.*				
Operations Guide	*Proj. Mgr.*				
Technical Test Plan	*Proj. Mgr.*				

ACTIVITY WORKSHEET IN WORK BREAKDOWN STRUCTURE DICTIONARY FORM

This is the author's favorite tool. The tool is referred to as the activity worksheet, but it is equally appropriate for describing any work at the process, activity, and task levels of a work decomposition. The information captured and maintained within a completed activity worksheet is reusable and provides the information that is necessary to build a WBS and project schedule.

One of the issues with the traditional waterfall based SDLC model is that it's very process centric and the time and effort to support an SDLC process maturity program can become quite burdensome. Anyone who has participated in a formal software process maturity program, such as CMMI, will know what I'm talking about.

But there's a better way to go about this activity. The activity worksheet provides a simple tool to employ, on an ad hoc basis, to capture the best information available within the organization on any given process, activity, or task. Specifically, as a project team encounters work that's not well-defined, ask a relevant Subject Matter Expert (SME) to document their approach to completing the scope of work outlined within the activity worksheet's work description.

Then, as the work is performed, the project team members should track actual data obtained from the execution of the work and update any variances they observe in the data. If the organization maintains an SDLC process repository, the worksheets can be used and modified across any number of projects. The activity worksheet provides a relatively simple and efficient approach to implementing an SDLC process improvement and maturity program, without all the burden and overhead of the commercial software process maturity programs.

The Activity Worksheet in Work Breakdown Structure (WBS) Dictionary Form contains the following data fields:

- Assigned process, activity or task identification number
- Date issued
- Assigned process, activity or task owner
- High level task description
- List of work activities
- Goals and objectives
- Identification of task deliverables
- Acceptance criteria
- Assumptions and constraints
- Skills and resources assigned to task(s)
- Materials required
- Estimated and actual task duration
- Estimated and actual task costs
- Estimated and actual due date
- Before and after task interdependencies
- Reviewers (or Approvals, if necessary)

YOUR COMPANY LOGO

Activity Worksheet in Work Breakdown Structure Dictionary Form
Project Name
Version

The WBS Activity Worksheet is made available to Subject Matter Experts (SMEs) to define the scope of work required for each activity and task within the work breakdown structure. For the entries made in this worksheet, accurate and activity and task descriptions can be compiled and tracked for variance during the course of a project.

Activity Worksheet in Work Breakdown Structure Dictionary Form							
Task Identification							
Task No.		Date Issued		Owner		Revision	
High Level Task Description	*Define the task*						
List Work / Activities	*What work is authorized?*						
Goals & Objectives	*Why is this work to be authorized?*						
Task Deliverables(s)	*Define the end product, service or result.*						
Acceptance Criteria	*What does "good" look like?*						
Assumptions	*What is required to get the work done?*						
Skills / Resources Assigned	*What kind or people and skills are required?*						
Materials Required							
Task Duration	Estimated		Actual				
Task Costs	Estimated		Actual				
Due Date	Estimated		Actual				
Task Inter-dependencies	Before Task						
	After Task						
Approved by	Project Manager		Date				

WORK BREAKDOWN STRUCTURE (WBS) RESOURCE PLANNING TEMPLATE

The WBS Resource Planning template includes a list of more than 200 common activities and tasks performed over the life cycle of a product, in a WBS format. However, unlike a project schedule plan, the WBS resource planning template provides columns to input the types of skills or resources required to support a project or program in lieu of scheduling data. The project manager or product owner, with input from the development team, inputs estimated hours by skills or resources as required for each defined activity and task.

The worksheet then calculates summaries of hours at the activity level, the phase level, and the project or program level. Value of this worksheet is that it provides an effective tool to estimate resource hours over the course of a program or project. Moreover, if the organization has a SDLC process maturity program, or uses the activity worksheets described in this book, the estimates will be more accurate.

The WBS Resource Planning Template is implemented as electronic worksheet and provides a matrix view of WBS tasks with the estimated duration of each task in hours with % of time required by the various skill sets to contribute to the tasks, summarized by total hours required for those skill sets. The worksheet includes a column for estimated duration and percent allocations are assigned for each required resource across each activity and task. The total hours summaries are automatically calculated.

The skill sets included in the worksheet are listed below, but users are free to add any other skill sets the required for their programs and projects:

- Project Manager
- Business Analysts
- Architect
- GUI developer
- Software development
- Technical Consultant
- Database Developer
- Quality Assurance / Testers

The breakdown template includes the following work categories:

- Project WBS
- Business Process Analysis
- System Design
- Commercial Off-The-Shelf (COTS) Product Assessments
- Detail Software & Hardware Configurations
- System & Hardware Implementation
- Custom Software Development
- Integration & Testing
- Deployment.

Work Breakdown Structure Resource Planning Template
Project Name
Version

YOUR COMPANY LOGO

The Work Breakdown Structure Resource Planning Template provides a matrix of WBS tasks with the estimated duration of each task in hours with % of time required by the various skillsets to contribute to the tasks, summarized by total hours required for those skillsets.

WBS No	Task Name	Estimated Duration	PM (%)	BA (%)	Architect (%)	GUI Dev (%)	Tech Cons (%)	SW Dev (%)	DB DEV (%)	QA/ Test (%)	Total Hrs
0	**Project WBS**	5	10%	100%	33%		33%				8.8
1	**Project Management**										0
1.1.	*Project Management Plan*										*0*
1.1.1.	Develop Project Charter										0
1.1.2.	Define Scope										0
1.1.3.	Define Resource Plan										0
1.1.4.	Develop Communications Plan										0
1.1.5.	Develop Risk Plan										0
1.1.6.	Develop Change Control Plan										0
1.1.7.	Develop Quality Plan										0
1.1.8.	Develop Purchase Plan										0
1.1.9.	Develop Cost Plan										0
1.1.10.	Develop Organization Plan										0
1.1.11.	Develop Project Schedule										0
1.2.	*Project Meetings*										*0*
1.2.1.	Conduct Kickoff Meeting										0
1.2.2.	Weekly Status Meetings										0
1.2.3.	Monthly Tactical Meetings										0
1.2.4.	Project Closing Meeting										0
1.3.	Administration										0
1.3.1.	Document Performance Standards										0
1.3.2.	Document Reporting Standards										0
1.3.3.	Document Naming Conventions										0
1.3.4.	Document Housekeeping Standards										0
	Sum Total =>	5	10%	100%	33%	0%	33%	0%	0%	0%	8.8
2	**Business Process Analysis (For Workflow-based Applications)**										0
2.1.	*Cross-Organizational Steering Committee*										*0*

www.SDLCforms.com

WORK BREAKDOWN STRUCTURE (WBS) TEMPLATE

A work breakdown structure provides a method that is used to decompose higher-level program and project deliverables into smaller deliverable components. The process of decomposition continues until work is defined at the lowest practical level. Work is defined at the lowest level of each branch within the WBS decompositions. The *8/80 rule* is often used as guiding principle to limit the size of WBS work tasks (a.k.a. work packages) to a range of hours between eight and 80 hours.

The WBS template is a simple form that the development team and project manager use to record the activities and tasks that must be performed over the course the project's duration. The decompositions are shown using a multilevel numbering list structure. E.g., 1.0, 1.1, 1.11, etc.

A work breakdown structure deliverable may be a product, data, service, or any combination thereof. In the traditional model, the WBS hierarchical structure along with information from the activity worksheets or other process descriptions provide the information necessary to define the project schedule.

However, it's important to understand that the WBS hierarchy does not describe the workflow necessary to build higher-level deliverables. Workflow is described by the inputs and outputs that are detailed in the process, activity, and task descriptions. Activity network diagrams help visualize the project's workflow. Gantt charts include the activity network diagrams or WBS with the dependencies, resources, start and finish dates, and task duration estimates to define the project schedule.

A WBS also provides the necessary framework to enable detailed cost estimating by identifying the major deliverables, and their underlying components and related work tasks. But again, cost estimating cannot occur at the WBS level because some activities and tasks may recur in the project's schedule. Most scheduling tools have capabilities to roll up costs at the activity and deliverable levels.

The Work Breakdown Structure contains the following data fields:

- Work Breakdown Structure:
  - Project Name
  - Department
  - Preparer
  - Project Number
  - Project Manager
  - Task #
  - Task Name
  - Description
  - Group or Individual Responsible
  - Due By

YOUR COMPANY LOGO

Work Breakdown Structure
Project Name
Version

Project Name	Department	Preparer	Date
Project Number	**Customer Contact**	**Project Manager**	

The Work Breakdown Structure table includes the tasks to be completed within a small project. Consider using a Project Management tool for large projects.

Task #	Task Name	Description	Group or Individual Responsible	Due by

COBIT CHECKLIST AND REVIEW TEMPLATE

After a series of highly visible cases of fraud and related business failures, the US Congress installed the Sarbanes-Oxley Act (2002) to mandate the type of information that all publicly held companies must report to show compliance with government and securities regulations. The scope of the SOX legislation covers corporate governance, risk management, auditing, and financial reporting of public companies. This legislation impacts business financial applications and other business systems that support those legislated activities.

There are various data tracking and reporting methods that have evolved to support the SOX act. One of the most rigorous frameworks supporting SOX is COBIT (Control Objectives for Information and Related Technologies), which is developed and supported by international professional association ISACA (Information Systems Audit and Control Association) for information technology (IT) management and IT governance.

COBIT provides management and business process owners with an Information Technology control model that helps the business understand and manage the risks related with IT. COBIT 5 is the current release, at the time this book was written, and is stated to provide a "*business framework for the governance and management of enterprise IT.*"

The COBIT Checklist and Review template is a tool that implements COBIT's rigorous structure to support Information Technology (IT) governance, accounting controls and compliance. In particular, the COBIT Control Objectives provides detailed objectives related to each IT process.

COBIT Control Objectives focuses on specific, detailed control objectives related with each IT process. For each of the 30+ IT structure processes, there are detailed control objectives that align the overall structure with objectives from primary sources comprising standards and regulations relating to IT. COBIT also includes statements of the desired results or objectives to be achieved by implementing the specified control procedures within each IT activity.

The COBIT Checklist and Review template contains the following topics:

- **Introduction**
- **COBIT Control Objectives**
- **COBIT Component Summary**
- **COBIT Processes**
  - Planning and Organization
  - Acquisition and Implementation
  - Delivery and Support
  - Monitoring

3 COBIT Component Summary

COBIT (Control Objectives for Information and Related Technology) is a complete structure for managing Information Technology (IT) risk and control. It includes four domains, 30+ IT processes, and 300+ detailed control objectives. It includes controls that address operational and compliance objectives.

Domain	Process Topics
Plan and Organize (IT Environment)	IT Strategic Planning Information Architecture Determine Technological Direction IT Organization and Relationships Manage the IT Investment Communication of Management Aims and Direction Management of Human Resources Compliance of External Requirements Assessment of Risks Manage Projects Management of Quality
Acquire and Implement (Program Development and Program Change)	Identify Automated Solutions Acquire or Develop Application Software Acquire Technology Infrastructure Develop and Maintain Policies and Procedures Install and Test Application Software and Technology Infrastructure Manage Changes
Deliver and Support (Computer Operations and Access to Programs and Data)	Define and Manage Service Levels Manage Third-Party Services Manage Performance and Capacity Ensure Continuous Service Ensure Systems Security Identify and Allocate Costs Educate and Train Users Assist and Advise Customers Manage the Configurations Manage Problems and Incidents Manage Data Manage Facilities Manage Operations

COBIT OBJECTIVES AND AUDIT ACTIVITY REPORT

The Sarbanes-Oxley Act, including COBIT Checklist and Review, provides for a standardized structure for Information Technology (IT) governance, accounting controls and compliance. COBIT Control Objectives focus on specific, detailed objectives related with each IT process.

The Audit Objectives and Audit Activity report provides a method and a tool to document applicable COBIT objectives and the status of each during periodic audits. The COBIT audit activity report provides a table structure to record the following information

- Control Objective ID
- Control Objective Name
- Control Activity ID
- Control Activity Name
- Applicable to SOX
- Status (complete, pending, incomplete)
- Description of the Control Activity.

COBIT Objectives and Audit Activity Report
Project Name
Version

Note: Text displayed in blue italics is included to provide guidance to the author and should be deleted before publishing the document. In any table, select and delete any blue line text; then click Home→Styles and select "Table Text" to restore the cells to the default value.

1 Purpose

This document provides a centralized source for defining and reporting on COBIT objectives and related activities.

2 Audit Objectives and Status

Control Objective ID	Control Objective Name	Control Activity ID	Control Activity Name	SOX Activity	Status	Control Activity Description
P04.11	*Segregation of Duties*	*P04.11.1*	*Segregation of Duties*	✓	*Complete*	*Roles are defined to ensure proper segregation of duty and to communicate the responsibilities that are aligned with that role.*
P04.11	*Segregation of Duties*	*P04.11.2*	*Segregation of Duties*	✓	*Incomplete*	*Authorization and roles are properly defined within IT systems to ensure segregation of duties, where possible the company uses logical and authorized mechanisms to enforce segregation of duties.*
P04.14	*Contracted Staff Policies and Procedures*	*P04.14.1*	*Contracted Staff*	✓	*Complete*	*Contractor and no-employee staff access, authorization, and accounts are managed in a consistent manner with employees.*
P04.9	*Data and System Ownership*	*P04.9.1*	*Data & System Ownership*	✓	*Incomplete*	*Applications have formal business owners to ensure accountability and authorization for changes to production environment.*

REQUEST FOR INFORMATION (RFI) TEMPLATE

A Request for Information (RFI) describes business problems or opportunities the customer hopes can be addressed by contractors, professional service organizations, or product vendors. The RFI should also include descriptions of desired capabilities that they need to address their problems and opportunities and specific deliverable items they require

An RFI is used in the information gathering phase, which in the traditional model is typically referred to as the concept phase. Vendor responses to the RFI help the business with its feasibility studies and cost-benefit analysis. However, data from the RFI is not typically used in the final business case. After the RFI responses are returned from the vendors, and the information is analyzed, a decision must be made on how to proceed. If the decision is to procure a vendor provided solution, the organization typically sends out a formal request for quote (RFQ). Where the RFI is an information gathering document, the RFQ requests written commitments from vendors in terms of deliverables, costs and schedules.

In addition to gathering basic information, an RFI is often used communicate the business opportunity to a broader audience of potential suppliers. The organization may already have ideas on which vendors can best fulfill their needs. However, it's not unusual for an unexpected vendor to step up and provide a unique and compelling solution for consideration.

The Request for Information contains the following topics:

- **Introduction and Purpose**
  - RFI Information
Confidentiality Information
  - RFI Process Stipulations
  - RFI Schedule
  - Vendor Presentation
- **Business Opportunity Overview**
  - Company Profile
  - Business Driver
- **Business Application Functions and Features**
- **RFI Response Configuration**
- **Vendor Response Information**
  - General Vendor Information
  - Vendor Vision and Capability
- **Product or Application Technical Information**
  - Product or Application Technical Functions & Features
  - Hardware Requirements
  - Software Requirements
- **Vendor Pricing Information**
- **Total Lifecycle Costs**

Request For Information
Project Name
Version

3 Business Opportunity Overview

3.1 Company Profile

Provide a description of <Company Name>.

3.2 Business Driver

Provide a description why your company is requesting information, i.e., high-level business drivers, e.g.,

- *Automate a specific function.*
- *Time delays in processing.*
- *Increase sales by x%.*

3.3 Business Application Functions and Features

Provide sufficient information concerning the specific business need for the vendors to understand the products, features, or services their response should be based on. This section is optional if the RFI is issued to obtain general company or product information.

Please fill in the following table and use a separate sheet of paper if necessary.

Business Application and Features	
Purpose	
Overview	
Objectives	
Functions	
Features	
Specifications	
Comparison to other competitors	

 Document: 2750 Page 8 of 23

www.SDLCforms.com

ROOT CAUSE ANALYSIS TEMPLATE

During a risk assessment it's very easy to key in on the outcomes or results of a problem and not the source of a problem. Addressing the outcome of the problem is a lot like putting lipstick on a pig. It may look prettier, but it doesn't really solve the underlying problem. To get to the source of a problem the organization may need to employ root cause analysis techniques.

Cause analysis provides a systematic method to identify and correct the root causes of events, rather than to simply address the symptomatic results. Implementation of corrective actions to address root causes has the added benefit of preventing recurrences of the problems.

The root cause analysis template helps the risk assessors evaluate the root cause of a problem, and then develop viable recommendations for a solution. The template also maintains information on the date the problem was encountered, the summary description of the problem, the duration of the problem, a list of the impacted business units and applications, and the recommended course of action to fix the problem, and any follow-up activities to ensure the fix solved the problem.

The Root Cause Analysis typically contains the following data fields:

- **Problem Summary Description**
  - Date Problem Occurred
  - Summary of Problem
  - Duration of Problem
  - Impact of Problem
  - Business Units / Sites / Departments Impacted
  - Applications(s) Impacted
  - Sequence of Events
  - Recommended Action Items.

YOUR COMPANY LOGO

Root Cause Analysis
Project Name
Version

Note: Text displayed in blue italics is included to provide guidance to the author and should be deleted before publishing the document. In any table, select and delete any blue line text; then click Home→Styles and select "Table Text" to restore the cells to the default value.

A **root cause** is a factor that caused a nonconformance and should be permanently eliminated through process improvement. **Root cause analysis** is a collective term that describes a wide range of approaches, tools, and techniques used to uncover **causes** of problems.

The Root Cause Analysis Template provides a template to define the approach taken to identify the cause, perform an analysis, and provide follow-up actions to address the root cause of a particular problem.

Date Problem Occurred	*99/99/9999*					
Summary of Problem	*Describe the problem or symptoms.*					
Duration of Problem	**Start Time**		**End Time**		**Duration (days, weeks, etc.)**	
Impact of Problem	*Describe the business / technical impact created by this problem.*					

Business Units / Sites / Departments Impacted	Application(s) Impacted
Describe who or what departments were impacted by the problem.	*Describe what application(s) were impacted.*

Sequence of Events:

Provide the sequence of events from the time the problem occurred to the time it was resolved.

Date	Time	Who	What
99/99/99	*10:00 AM*		

www.SDLCforms.com

PROJECT PLAN SCHEDULE TEMPLATE

The project's schedule plan lists the activities and work tasks necessary to complete the total approved for the project. The project plan schedule template should list most of the common activities and tasks across the entire software and systems development lifecycle. When the project scope involves only a subset of the entire SDLC, the project manager simply removes the activities and work activities that do not apply to their project requirements.

A project schedule often looks like the WBS applied to a timeline. However, developers of the schedule plan need to be careful because a WBS does not describe when multiple deliverables of the same type are required, or when specific work activities may occur multiple times within the project. Those details are worked out during the construction of network activity diagrams and Gantt charts.

The schedule plan includes additional information, typically implemented in forms or columnar formats, or both, describing activity and task estimated durations, resources, start and finish dates, percent complete, and other useful information. The Project Plan template shown on the next page implements common data fields required to build a project schedule and copy it into a project scheduling tool.

The project plan is created by the project manager and should include input from the appropriate activity worksheets. Qualified team members who must perform the work are always the best source of task related information and estimates. They should complete the activity worksheets for the work thy are assigned.

The project plan schedule template described in this book follows the traditional SDLC phases and includes data fields to capture the following information:

- **Task**
  - Task Name
  - Duration
  - Percent Complete
  - Priority
  - Start Date
  - Finish Date
- **Predecessors**
  - ID
  - Task Name
  - Type Lag
- **Resources**
  - Resource Name
  - Units
- **Constraints**
  - Deadline
  - Constraint Type
  - Constraint Date
  - WBS Code
  - Earned Value Method

ID	Task Name	Duration	Start	Finish	% Complete	Resource Names
1	**XXX Project**	**1 day?**	**Mon 1/3/11**	**Mon 1/3/11**	0%	
2	**Project Initiation Phase**	**1 day?**	**Mon 1/3/11**	**Mon 1/3/11**	**0%**	
3	Project initiation XXXXXX Submitted and Approved	1 day?	Mon 1/3/11	Mon 1/3/11	0%	Project Manager
4	Incorporate Project in Company Accounting Systems	1 day?	Mon 1/3/11	Mon 1/3/11	0%	Project Manager
5	Define Project in Appropriate Company Site	1 day?	Mon 1/3/11	Mon 1/3/11	0%	Project Manager
6	Develop and Approve Business Case Document	1 day	Mon 1/3/11	Mon 1/3/11	0%	Business Project Manager,Technica
7	Obtain XXX Work Order for project	1 day	Mon 1/3/11	Mon 1/3/11	0%	Project Manager
8	Prepare Cost-Benefit Analysis of Project	1 day?	Mon 1/3/11	Mon 1/3/11	0%	Business Analyst
9	Prepare Capital vs. Expense Cost Worksheet	1 day?	Mon 1/3/11	Mon 1/3/11	0%	Project Manager,Business Project M
10	Prepare Initial Project Plan	1 day?	Mon 1/3/11	Mon 1/3/11	0%	Project Manager
11	Define Project Team	1 day?	Mon 1/3/11	Mon 1/3/11	0%	Project Manager
12	Create Project Checklist	1 day?	Mon 1/3/11	Mon 1/3/11	0%	Business Analyst
13	**Project Planning Phase**	**1 day?**	**Mon 1/3/11**	**Mon 1/3/11**	**0%**	
14	Develop and Approve Project Charter	1 day?	Mon 1/3/11	Mon 1/3/11	0%	Technical Writer[50%],Project Mana
15	Develop and Approve Statement of Work	1 day?	Mon 1/3/11	Mon 1/3/11	0%	Technical Writer[50%],Business An
16	Prepare Project Approval Document	1 day?	Mon 1/3/11	Mon 1/3/11	0%	Business Analyst
17	Obtain approval for Project Approval Document	1 day?	Mon 1/3/11	Mon 1/3/11	0%	Business Project Manager
18	Develop Project Capital vs. Expense Costs Worksheet	1 day?	Mon 1/3/11	Mon 1/3/11	0%	Project Manager
19	Prepare Project Risk Analysis	1 day?	Mon 1/3/11	Mon 1/3/11	0%	Technical Writer,Project Manager
20	Prepare Procurement Plan	1 day?	Mon 1/3/11	Mon 1/3/11	0%	Technical Writer,Business Analyst
21	Develop Roles and Responsibilities Matrix	1 day?	Mon 1/3/11	Mon 1/3/11	0%	Project Manager
22	Develop Required Approvals Matrix	1 day?	Mon 1/3/11	Mon 1/3/11	0%	Project Manager
23	Obtain approval for Required Approvals Matrix	1 day?	Mon 1/3/11	Mon 1/3/11	0%	Business Project Manager
24	Prepare RFP/RFQs for "vendor name"	1 day?	Mon 1/3/11	Mon 1/3/11	0%	Project Manager,Technical Writer,B
25	Refine and obtain approval for Project Plan	0.5 days?	Mon 1/3/11	Mon 1/3/11	0%	Business Project Manager,Project M
26	**Requirements Definition Phase**	**1 day?**	**Mon 1/3/11**	**Mon 1/3/11**	**0%**	
27	Develop Requirements Checklist	1 day?	Mon 1/3/11	Mon 1/3/11	0%	Technical Writer,Project Manager
28	Develop Business Requirements Document	1 day?	Mon 1/3/11	Mon 1/3/11	0%	Technical Writer,Business Analyst
29	Obtain approval for Business Requirements Document	1 day?	Mon 1/3/11	Mon 1/3/11	0%	Business Project Manager
30	Develop Functional Requirements Document	1 day?	Mon 1/3/11	Mon 1/3/11	0%	Technical Writer,Business Analyst,E
31	Obtain approval for Functional Requirements Document	1 day?	Mon 1/3/11	Mon 1/3/11	0%	Business Project Manager
32	Finalize vendor contracts	1 day?	Mon 1/3/11	Mon 1/3/11	0%	Business Analyst
33	Procure vendor items	1 day?	Mon 1/3/11	Mon 1/3/11	0%	Procurement

T W T F

Project: Project Plan
Date: Wed 11/4/15

Task	Inactive Task	Manual Summary
Split	Inactive Milestone	Start-only
Milestone	Inactive Milestone	Finish-only
Summary	Inactive Summary	External Tasks
Project Summary	Manual Task	External Milestone
External Tasks	Duration-only	Progress
External Milestone	Manual Summary Rollup	Deadline

www.SDLCforms.com

PROJECT MANAGER'S WEEKLY CHECKLIST

The project manager's weekly checklist provides a list of weekly tasks performed by IT project managers to remind them of critical tasks to keep the project running smoothly. These tasks can be listed in the project management section within the project schedule, however it's often easier to manage the tasks in a separate and simpler checklist.

The project manager's weekly checklists separate tasks into the following categories:

- Project Start - Tasks to be completed in the first few weeks including verifying that the project charter/business case has been developed, that the project objectives have been clearly developed, the project data sources are clearly understood and available, that a project has been developed, etc.
- Mid Project - Tasks to be completed every week thereafter including a frequent review of the project scope, risks and issues, that no changes are necessary to the project budget and schedule, that project sponsors and stakeholders are continually being updated with comprehensive status reports, etc.
- Project Close - Tasks to be completed for project closure including has the application been delivered to production, has the project sponsor accepted the final project, are disaster recovery procedures in place in the event of catastrophic failure, etc.?

Project managers can add tasks to the checklist. And there is no reason why program managers, product owners, and other project team members can't build their own checklists.

Project Manager's Weekly Checklist
Project Name
Version

To ensure your project stays on track, review each item below on a weekly basis.

Week Ending:	Monday, January 31, 2019

	First Few Weeks	Date Completed
☐	**Project Charter -** Has a project charter been developed? If not, immediately create one.	
☐	**Business Case -** Has a Business Case been reviewed and approved?	
☐	**Project Purpose –** Has the project purpose been communicated to all team members and stakeholders?	
☐	**Project Objectives -** Identify project objectives if these are not included in the Business Case document.	
☐	**Project Scope -** Identify everything that is to be included in a project scope if these items have not been in the Business Case document. These items will be explained in greater detail in the Business Requirements Document.	
☐	**Project Budget -** Verify that the project budget (consisting of both capital and expense items) has been established to provide sufficient monies to cover all expenses plus a contingency fund of at least 10%. Ensure that the budget has been approved by all stakeholders.	
☐	**Project Data Sources –** Verify that all project data and control sources (General Ledger accounts, work process flow, project dependencies, change control process, project management software, etc.) that will be required to determine project status have been identified and are available for daily use. Refer to the Project Management Office (PMO) Checklist for a list of these items.	
☐	**Project Risks –** Identify potential project risks, and categorize each as to the degree of severity.	
☐	**Project Issues –** Identify potential project issues.	
☐	**Work Breakdown Structure –** Has the Work Breakdown Structure been developed?	
	Project Plan – Has a project plan been developed precisely identifying the task breakdown by role to reflect the Work Breakdown Structure?	
☐	**Project Organization Chart –** Has a Project Organization of all stakeholders and project team members been created and distributed to all departments?	
☐	**Roles and Responsibilities –** Has a Roles and Responsibilities Matrix been developed and discussed with all project team members?	
☐	**Project Team Members –** Identify all project team members and their availability for the project – part-time or full-time.	
☐	**Setup User Accounts –** Setup the user accounts for your project management software and provide user instructions or classes in order that everyone can use the software.	
☐	**List of Deliverables –** Generate an initial list of deliverables for the project.	

LIST OF OPPORTUNITIES SUMMARY TEMPLATE

The List of Opportunities Summary is a tool that summarizes identified project opportunities, and includes data fields to document the opportunity's description, priority, target date for delivery, and owner. The form documents opportunities in form of deliverables and tasks not identified in the project schedule.

Opportunities are items that have not risen to a level of priority to justify a formal change request and analysis. They represent future possibilities. Opportunities are also not action items. Action items are previously unidentified work tasks, discovered on an ad hoc basis, that must be accomplished.

The List of Opportunities Summary contains the following data fields:

- **Opportunity**
  - Topic / Phase
  - Opportunity Description
  - Priority (High/Medium/Low)
  - Target Date
  - Actual Date
  - Owner
  - Comments

YOUR COMPANY LOGO

List of Opportunities Summary
Project Name
Version

Note: Text displayed in blue italics is included to provide guidance to the author and should be deleted before publishing the document. In any table, select and delete any blue line text; then click Home→Styles and select "Table Text" to restore the cells to the default value.

Sample Topics or Phases: Initiation, Analysis, Requirements, Specifications, Design, Build, Test (e.g., unit, system, integration, regression, UAT), Documentation, Training, Implementation, Roll-Out, Decommission

Other: Change Control, Communication, Compliance, Cost Management, General Process, Human Resources, Issue Management, Post-Implementation, Risk Management

Topic / Phase	Opportunity Description	Priority (H/M/L)	Target Date	Actual Date	Owner	Comments

Requirements Phase
Overview

The methods and tools associated with the requirements phase facilitate the work required, and help identify and document business, customer, and user requirements. Note that many of the analytical tools and checklists provided in this section have viability even under agile practices, as they provide a list of things to check to confirm the thoroughness of the requirements analysis.

Requirements analysis provides a deeper dive into business and user needs, when compared to the feasibility studies and cost benefit analysis efforts described in the concept phase. The focus of requirements analysis is on understanding the issues that managers, process owners and end-users face in their work, and seeking better process outcomes, less waste, and more efficiencies.

Business and user requirements are sometimes referred to as functional requirements, the implication being the requirements detail the functionality required by the software system. However, there is a necessary counterpart to functional requirements that must also be captured, which are the non-functional requirements.

Non-functional requirements are the technical implementation capabilities the software systems must support. Non-functional requirements are sometimes referred to as the "*abilities.*" E.g., terms and metrics that specify the desired or target operational characteristics of a system, such as accessibility, extensibility, reliability, maintainability, testability, maintainability, scalability, safety, security, and usability.

The development team determines the non-functional requirements via the metrics and performance capabilities specified in the business and user functional requirements. In other words, whether the requirements analysis work is assigned to business analysts or development team members, business and user requirements must include details on acceptance criteria and the performance measures expected by users and customers.

Business and user requirements define requirements in the form of desired capabilities and not in the form of product features and functions. Features and functions, or modifications to existing products, are implementation details that are sorted out in the design phase of work.

Business systems typically support cross-functional and cross-organizational business processes. When defining business processes, the business analysts should simultaneously evaluate the efficiency and capabilities of the existing processes before defining the requirements for a new system or enhancements to an existing business system.

This is an important concept to understand. There is a real danger to automating business processes that are not efficient or streamlined, and wasteful, and therefore not adding value to the customer. Automating an existing process may have the opposite effect that was intended, by faithfully executing the codified process workflows and thereby increasing non-value-added work, work in process, operational complexity, and costs.

Often times, in order to analyze the future needs of the organization, it's useful to first document how things are currently done today. This type of requirements analysis is referred to as the '*current state*' or *as-is* assessment. The as-is assessment helps the business analyst not only document how things are currently done, but also delve deeper into understanding how and where things are not working very well.

Next, the business analysts work with stakeholders, end-users and customers to-determine to determine an ideal "future state." The future state (a.k.a. "to-be" state) analysis evaluates three aspects of a business process:

1. Define activities, data, and human and system interactions necessary to support the process
2. Eliminate existing activities that do not add value to the customer
3. Determine what system capabilities are required to improve process outcomes, throughput, and efficiencies

The ideal goal of the combined as-is assessments and to-be analysis is to understand how the business can remove waste in the form of non-value-added activities and be more customer centric - before figuring out how to automate the affected business processes.

Once the as-is assessment and the to-be analysis are complete, the business analyst will turn their attention to developing a "gap analysis." This is an important step that determines the magnitude of change that is required to move the business as a complex system from how things are done today to how they need to be done in the future.

During the requirements analysis, bear in mind that business systems facilitate proper and complete execution of end-to-end workflows across critical business processes. People and processes are not the only enablers within a business. The software systems and enhancements can enable positive effects on a business process, by providing process and task performance support capabilities to the end-users, and by making the overall process more efficient, adding new functional capabilities, and by providing accurate information on timely basis.

The business requirements for the implementation of new features and functions may require changes to the underlying IT infrastructure, including servers, networks, storage devices, and security. Since business systems often change business processes, the skills and roles of the organization's employees may have to change. And, for more complex systems, it's possible that new equipment and technologies are also involved to implement the new business requirements.

One final note: in my book titled *Frameworks for Agility*, I reference the *Business Analysis Body of Knowledge® Guide (BABOK® Guide), published by the* International Institute of Business Analysis (IIBA) as a primary source of business analysis standards.[6] The BABOK Guide identifies numerous techniques and tools supporting business analysis. The templates provided within this book represent a small subset of potential methods to evaluate business and user needs and activities.

Forms & Templates

MANAGING SCOPE AND REQUIREMENTS CHECKLIST

Under the traditional software development model, plans are everything. And, one of the biggest ways to fail under the traditional model is to allow scope creep to set in. Scope creep simply means the project has taken on more work than was previously approved, which drives up the cost by adding work, resource requirements, and time delays to the schedule.

Managing the project scope, requirements, and meeting customer expectations are major factors in the success of a project. Poor scope definition, stakeholder analysis, and weak sponsorship could lead to costly and untimely project failure.

This checklist helps you to manage project scope and requirements, e.g.,

- Identify project stakeholders
- Ensure the requirements are documented
- Obtain commitment on deliverables, schedules, and resources
- Handle changes to the project scope
- Acquire early agreement on project scope to avoid scope creep
- Foresee and manage risks
- Attain greater customer satisfaction and effective execution.

However, I would argue that scope verification and approvals are also important factors under agile-based software development models. Paying customers and executive sponsors have constraints. They expect results within the schedule, and with the resources and budgets allotted to the project. Customers can quit funding agile projects just as quickly as they can on traditional projects - if and when they think the development team is not focused correctly on the work they have commissioned.

A product owner can build an extensive product backlog, but those requirements must map back to the customer's authorized scope of work. and delivered within the constraints that justified the investments. If the project or product team does not have a formal agreement or project charter that explicitly identifies the approved scope of work and expected deliverables, the *Managing Scope and Requirements Checklist* is a useful tool to document the scope of work expected by the paying customers or executive sponsors.

The Managing Scope and Requirements Checklist contains the following topics:

- Purpose
- Product or system overview
- Reason for application, system or project
- Assumptions, dependencies, and constraints
- List of stakeholders
- Risks
- Scope/Requirements checklist.

YOUR COMPANY LOGO

Managing Scope and Requirements
Project Name
Version

3 Checklist for Managing Scope and Requirements

Fill in the following table with the appropriate information.

Topic	Yes	No	Comments
Customer agrees with the scope of the project.			
Project scope has been documented and is measurable.			
The scope statement provides a meaningful baseline for ongoing management (i.e., it is clear when proposed changes are in or out of scope).			
The customer understands the importance of scope management?			
The scope statement been signed-off by the Customer and other Stakeholders and formally baselined.			
The scope is relatively stable.			
Confirmation has been obtained that commitments must be renegotiated and project work plans revised, as appropriate, whenever the approved scope changes.			
The project will maintain a profile of the software/hardware size over time for configuration items.			
Customer agrees with the requirements that will be addressed in the project.			
The requirements are clearly defined.			
Project includes a traceability matrix to show where each requirement is formally addressed.			
All requirements have been signed-off by the Customer and Stakeholders and formally baselined.			
Measurements are used to determine the status of the requirements (e.g., total number of requirements, changes that are proposed, open, approved, and included into the baseline).			
The requirements are relatively stable.			
Confirmation has been obtained that commitments, project plans, work products, and activities must be			

SIMPLE APPROACH TO REQUIREMENTS AND SYSTEMS ANALYSIS TEMPLATE

Smaller and less complex development efforts do not require the same level of business, user, and systems requirements analysis and documentation. The Simple Approach to Requirements and Systems Analysis template offers a simpler tool to document the identified business, user, and system requirements.

Topics covered within the Simple Approach to Requirements and Systems Analysis template include:

- Problems or issues to resolve, objectives or goals met with this solution, and the solution to be implemented
- Why the solution is being implemented
- Project related information, description, approach, goals, objectives and scope
- Type Software Development Life Cycle (SDLC) process employed
- Project Specifications, schedule and location considerations and system specifications
- An Objective Considerations Checklist
- Task assignment and project resources
- A roles and responsibilities matrix
- Staffing needs and costs
- High-level business requirements, system interfaces and infrastructure requirements
- Usability, performance, operations and maintenance requirements
- Data, sample reports, and screen requirements
- Training and documentation requirements
- Design assumptions and support considerations
- Architecture and interface design, application layer information and implementation and software execution

YOUR COMPANY LOGO

Requirements and Systems Analysis
Project Name
Version

4 Project Specifications

4.1 Scope and Limitations

Topic	Content
General Area to be Studied	
Contact Name, Title, and Department	
Location and Phone	
Email Address	
Specific Functions to be Covered in Analysis	
Specific Project Constraints/Limitations	

4.2 Schedule and Location Considerations

Topic	Content
Schedule Factors:	
Project Start Considerations	
Project Completion Considerations	
Study Area Geographical Location(s)	
If Travel is Required, specify	
Will Client Provide Work Space, if yes, specify	

4.3 System Specifications

Topic	Content
Name of System	
List Current Problem(s)	
List Prime Objective(s)/Goal(s), see checklist	
Describe Specific System Constraints/Limitations	
Briefly Describe Existing Systems (e.g., programs, servers, databases, technology and other equipment)	

BUSINESS REQUIREMENTS DOCUMENT (BRD) TEMPLATE

A Business Requirements Document (BRD) is a formal document that implements a rigorous process to capture business requirements. The breadth of a BRD encompasses the general business requirements, including the business and end user requirements, business problems or issues, project information, process information, training, and documentation requirements.

The most common objectives of the Business Requirements Document (BRD) are to:

- **Document and gain stakeholder approval on identified business requirements and priorities**
- **Communicate how a proposed solution satisfies business and user needs**
- **Provides input into the architecture and design requirements**

The BRD is a critical document within the traditional SDLC process, as it provides an expansive view that incorporates management, stakeholder, and end-user views on business requirements supported by the proposed software system. After the product is constructed, the BRD also provides detailed information on the capabilities implemented within the product.

A BRD typically contains the following sections and data fields, though it should be modified to meet the needs of each program or project:

- **Project Information**
  - Project Description
  - Project Approach
  - Goals, Objectives, and Scope
  - Business Drivers
  - Stakeholders
  - Assumptions, Dependencies, and Constraints
  - Risks
  - Costs
  - Delivery Dates
- **Process Information**
  - Current Processes
  - New Processes or Future Enhancements
- **Requirements Information**
  - High-Level Business Requirements
  - System Interfaces
  - Infrastructure Requirements
- **Other Requirements Information**
  - Product, Organizational, System, and External Requirements
  - Usability, Performance, Operations, and Maintenance Requirements
  - Content/Data/Sample Report Requirements
  - Screen Requirements
  - Training and Documentation Requirements

Business Requirements Document
Project Name
Version

2.3 Goals, Objectives, and Scope

The objectives statement describes the goals of the project. The Project Scope statement defines the scope of the project.

List the high-level project deliverables and specify whether they are included or excluded from the project.

ID	Description	Included/ Excluded

2.4 Business Drivers

Describe why the project is being implemented. This may be to reduce costs, increase revenue, streamline processes, etc.

2.5 Stakeholders

The Stakeholders statement and table outlines the project stakeholders and their respective roles.

Name	Department	Role
John Smith	*Accounting*	*Manager*

2.6 Assumptions, Dependencies, and Constraints

List assumptions, dependencies, and constraints that could constrain the project team, development, or implementation, e.g., system hardware or software availability or access.

2.7 Risks

Provide information about any risks associated with the project. Provide any work-around or mitigation information for each risk identified.

BUSINESS REQUIREMENTS PRESENTATION TO STAKEHOLDERS TEMPLATE

Business Requirements Documents are very detailed and often quite lengthy. As a result, it may be difficult to get key executives and stakeholders to take the time to read through all the information provided. The *Business Requirements Presentation to Stakeholders* template is a tool to document important information in a concise format that can be delivered during a customer review meeting.

The template allows the business analysts to create a slideshow, using a tool such as Microsoft's PowerPoint application, to review identified project business requirements with the stakeholders and business executives who are sponsoring the project.

The presentation slides contain the following sections and information topics:

- **Purpose**
- **Project information**
  - Project background
  - Project description
  - Project approach
- **Goals, objectives and scope**
- **Project stakeholders**
- **Assumptions, dependencies and constraints**
- **Project Costs**
  - Cost item
  - Project cost – capital or expense
  - Annual maintenance / administration / support costs
- **Milestone delivery dates**
- **Current and future process flows**
- **High level business requirements**
- **System interfaces**
- **Infrastructure requirements**
- **Other requirements.**

Project Name
Business Requirements
Presentation to <Business Units>
Presented by <Your Name, Title>
Date
CONFIDENTIAL
<Your company logo or name>
www.SDLCforms.com

FUNCTIONAL REQUIREMENTS DOCUMENT (FRD)

Under the traditional software development model, the functional requirements document takes the capability requirements described in the BRD and other requirements documentation and starts the assessment of how they are implemented within software product or system. The BRD typically includes descriptions of data inputs and outputs, screens and their functional purpose, descriptions of work and data flows, and desired reports, analytics, and data outputs.

An FRD may include descriptions of data calculations, technical details, data manipulation and processing, and other capability functionality that define *what a system is supposed to accomplish*. Functional requirements are supported by non-functional requirements (also known as *quality requirements* or *"abilities"*), which impose performance constraints on the design or implementation (such as reliability, extensibility, accessibility, security, and supportability).

An FRD is technology agnostic and lacks implementation details, having a focus on describing the general capabilities and behaviors of the proposed system. The *System Requirements Specification (a.k.a. Design Plan) describes the* details on how the system implements the functional requirements, while the plan for implementing *non-functional requirements is* described in the System *Architecture Document*.

The Functional Requirements Document: contains the following sections and topics:

- Purpose
  - Project Description
  - Project Approach
  - Goals, Objectives, and Scope
  - Business Drivers
  - Stakeholders
  - Assumptions, Dependencies, and Constraints
  - Risks
  - Costs
  - Delivery Dates
- Process Information
  - Current Processes
  - New Processes or Future Enhancements
- Requirements Information
  - Functional Requirements
  - Infrastructure Requirements
  - Other Requirements
  - Non-Functional Requirements
- Interfaces
  - System Interfaces
  - Software Interfaces
  - Hardware Interfaces

Functional Requirements Document
Project Name
Version

2.3 Goals, Objectives, and Scope

The objectives statement describes the goals of the project. The Project Scope statement defines the scope of the project.

List the high-level project deliverables and specify whether they are included or excluded from the project.

ID	Description	Included / Excluded

2.4 Business Drivers

Describe why the project is being implemented. This may be to reduce costs, increase revenue, streamline processes, etc.

2.5 Stakeholders

The Stakeholders statement and table outlines the project stakeholders and their respective roles.

Name	Department	Role
John Smith	*Accounting*	*Manager*

2.6 Assumptions, Dependencies, and Constraints

List assumptions, dependencies, and constraints that can constrain the project team, development, or implementation, e.g., system hardware or software availability or access.

 Document: 3150 Page 5 of 14

USE CASE TEMPLATE

The Use Case requirements modeling method supports the Unified Modeling Language (UML). UML is a general-purpose modeling language that evolved to support object-oriented software development approaches. A use case describes *"who" can do "what"* with a proposed system and describes the set of events or interactions as a series of steps between users and systems.

The use case method describes requirements as a set of interactions between an "actor" and a "system" to achieve a desired outcome or objective. An actor is a person or some other system that needs a service from the target system that provides the service.

The use case approach to requirements analysis helps the business analyst or developer analyze the business or user problems to be resolved. The use case technique is used to capture a system's behavioral requirements by detailing scenario-driven situations through the functional requirements and may evaluate more than one potential scenario. When completed, the use case documents the proposed solution to be implemented and describes why the solution is the best alternative.

The use case template described below, and as shown on the next page, features the following information fields to prepare a use case requirement document.

- **Purpose of the Use Case**
- **Project Information**
  - Project Approach
  - Goals, Objectives, and Scope
  - Business Drivers
  - Stakeholders
  - Assumptions, Dependencies, and Constraints
  - Risks
  - Costs
- **Process Information**
- **Requirements Information**
  - High-Level Business Requirements
  - System Interfaces
  - Infrastructure Requirements
- **Use Case Specifications**
  - Overview
  - Simplified Method and Traditional Method
  - Sample Template
  - Product, Organizational, System, and External Requirements
  - Usability, Performance, Operations and Maintenance Requirements
  - Content / Data / Sample Report Requirements
  - Screen Requirements
  - Training and Documentation Requirements

Use Case Template
Project Name
Version

Note: Text displayed in blue italics is included to provide guidance to the author and should be deleted before publishing the document. In any table, select and delete any blue line text; then click Home→Styles and select "Table Text" to restore the cells to the default value.

1 Purpose

Purpose describes the intent of the document, which is to define the business requirements for the project using a use case methodology.

- *Problems or issues to resolve.*
- *Objectives or goals met with this solution.*
- *Solution to be implemented.*
- *Why the solution is being implemented.*

2 Project Information

This section of the document should contain project background information including the project's objectives and major system functionality.

2.1 Project Description

The Project Description section provides a high level view of the project; background, vision, approach, timeframe, etc.

2.2 Project Approach

The Project Approach statement provides a complete description of the approach to be taken for the delivery of the project. Outline the phases of the project including the high-level activities. If applicable, describe how a specific phase relates to the overall project.

2.3 Goals, Objectives, and Scope

The objectives statement describes the goals of the project. The Project Scope statement defines the scope of the project.

List the high-level project deliverables and specify whether they are included or excluded from the project.

ID	Description	Included / Excluded

REQUIREMENTS INSPECTION CHECKLIST

This is a tool that has use within both agile and traditional-based software development practices. Although agile practitioners' frown on excessive documentation, there is no reason why the developers cannot use a tool such as this *Requirements Inspection Checklist* to review, assess and validate the thoroughness of their requirements gathering and analysis activities. In other words, they may not fill out and turn in the form, but simply use the checklists to quickly review the completeness of their work.

The Requirements Inspection Checklist is a simple quality assurance tool to verify at a glance that major requirement functions and tasks have been completed. The value of the tool is to ensure those who have documented the business, user and technical requirements have rigorously captures and properly documented the identified requirements, verified the correct interpretation of the requirements with their customers, tested calculations and algorithms that must be implemented in the software application or system, and initiated the development of the traceability matrix.

Over the course of the development activity, the traceability matrix helps ensure every requirement has one or more related test cases or test scenarios, code, test results, and verification of conformance. The Requirements Inspection Checklist helps set up the traceability process by ensuring completeness of the requirements gathering and documentation process. Errors and omissions can dramatically affect the timeliness, quality, and costs of the project.

The Requirements Inspection checklist contains the following sections with Yes/ No checkboxes for each identified checklist item:

- **Purpose**
  - Stated purpose of the requirements checklist
- **Requirements Checklist**
  - General Information
  - Correctness
  - Requirements Traceability
  - Interfaces
  - Behavioral requirements
  - Non-Behavioral requirements
  - Other Information.

YOUR COMPANY LOGO

Requirements Checklist
Project Name
Version

Note: Text displayed in blue italics is included to provide guidance to the author and should be deleted before publishing the document. In any table, select and delete any blue line text; then click Home→Styles and select "Table Text" to restore the cells to the default value.

1 Purpose

The purpose of the Requirements Inspection Checklist is to provide a sample quality assurance document to verify that major requirement functions and tasks have been completed.

2 Requirements Checklist

2.1 General Information

Requirement Information	Yes / No
All requirements prepared at a consistent level of detail.	
Requirements provide sufficient information for design.	
All cross-references to other requirements are correct.	
Priority for implementing each requirement is included and correct.	
Identified and addressed input value expected ranges for each scenario.	
Defined all external hardware, software, and communication interfaces.	
Defined functional requirement algorithms.	
Defined all functional requirement business rules.	
Software requirement specifications include all known customer or system needs.	
Performance requirements for timing, memory, and resource use are easily understood.	
Documented responses for valid and invalid input values.	
Documented error condition expected behavior.	
Requirements indicate what the system should do and not do.	
Standards are provided (where applicable),e.g., for transaction speed, database integrity, resource limits, operating environments, etc.	

REQUIREMENTS TRACEABILITY MATRIX (RTM) TEMPLATE

The RTM is a tool that helps verify the association of identified requirements, as documented within the BRD and other Requirements Specification – including design and testing documentation. The design documents describe how the capabilities defined within the requirements are implemented. The code implements the desired business, user and technical capabilities. The testing documents describe how the code is tested to ensure the code implements each requirement correctly. And, the testing results and compliance certification documents confirm the code's compliance, or not.

The term *requirements traceability* refers to the methods and tools used to follow work accomplished against each requirement to design, implement, test and confirm achievement of the desired capability and compliance with the acceptance criteria for each requirement.

Regardless of which technique is used to analyze and document requirements, the RTM helps ensure every requirement is properly designed, implemented, tested and confirmed, or fixed as necessary. The RTM just as useful when tracking completion of stories in the product backlog within an agile-based development project, as it is in the traditional SDLC approaches.

The Requirements Traceability Matrix document contains the following sections and topics:

- Purpose
- Requirements Checklist
  - General Information
  - Interfaces
  - Behavioral Requirements (e.g., What the system must do to support user interactions)/ Functional Requirements (e.g., business rules, process flows, calculations and analytics)
  - Non-Behavioral Requirements (e.g., Attributes of the system)/ Non-functional Requirements (e.g., the *"abilities")*
  - *Correctness/ Acceptance Criteria*
  - *Requirements Traceability Matrix (RTM)*
  - *Other Information*

Requirements Traceability Matrix
Project Name
Version

YOUR COMPANY LOGO

4 Traceability Matrix Samples

Companies require some type of verification that ensures all requested requirements and design information was actually tested and approved. Each company must decide how to document traceability, which may be based on corporate culture, management style, and project complexity.

Three sample matrix tables are shown for review and selection.

4.1 Simple Matrix

The following table provides a table structure for a simple traceability matrix when there is limited documentation to reference.

Req. #	Requirement	Test Case / Procedure	Test Date(s)	Status / Comments
	Key Requirements (section 1)			
1.	*Requirement 1*			
2.	*Requirement 2*			
3.	*Requirement 3 ......*			
	Other Requirements (section 2)			
4.	*Requirement 4*			
5.	*Requirement 5*			
6.	*Requirement 6 ......*			

 Document: 3350 Page 7 of 10

REQUIREMENTS CHANGES IMPACT ANALYSIS FORM

A requirements change impact analysis evaluates both business and technical considerations, as well as the costs and effort to implement proposed changes. The impact assessment includes identifying the requirements or specifications that will be modified and indicating any dependencies and relationships.

Under the traditional software development model, this form helps perform the impact analysis and document the results to make a formal change request. However, this is another tool that has use within both agile and traditional-based software development practices.

Although agile practitioners' frown on excessive documentation, there is no reason why the developers cannot use a tool such as this *Requirements Change Impact Analysis* form to review and assess the impact of proposed changes to a system under development. In other words, they may not fill out and turn in the form, but simply use the form to quickly review and understand the impact of the proposed changes in requirements.

The form provides a list of criteria that guide impact analysis of requirement changes, including proposed change implications, identification of affected system components, the identification of affected organizations, users and stakeholders, and estimated schedule and cost impacts.

This template helps with assessing the impact of change requests in the following areas:

- Business implications of the proposed change
- System elements affected by the proposed change
- Effort, schedule and cost impact estimations
- Impact analysis analytics and reporting.

The Requirements Changes Impact Analysis contains the following sections:

- Purpose
- Product or system information
- Reason and description of requirements changes
- Assumptions, dependencies, and constraints
- List of stakeholders
- Risks
- Proposed changes checklist
- System components and elements affected by change
- Estimated schedule and cost impact.

3 Requirement Change Impact Analysis

3.1 Proposed Change(s) Implications

Complete the following table to identify proposed change implications.

Topic	Description or Comments
List existing requirements that differ with the proposed change.	
List additional pending requirement changes that differ with the proposed change.	
What are the negative results of not making the change?	
What are possible unfavorable effects or risks of making the proposed change?	
Will the proposed change unfavorably affect performance or other feature or function attributes?	
Indicate whether the change will impact system modules or elements and affect critical features or functions (e.g., security).	
Will the change involve a product change or recertification?	
Can the proposed change be completed within technical limitations, restrictions, and current staff skills?	
Will the change place undesirable stress on any computer resources needed for the development, test, or operating environments?	
Does the change require additional tools to be acquired to execute and test the change?	
How will change affect the sequence, dependency, effort, or duration of tasks in the project plan?	
Will models, examples, samples or other user input be required to verify the change?	
How much project time and work has already been performed that will be lost if this change is accepted?	

TRAINING PLAN TEMPLATE

Training is often required when a new system or enhancements to an existing system change the organization's business processes or change the way users interact with an existing system. Besides user training requirements, there may be a need to train the support, maintenance, and other operations-oriented staff.

In a modern context, training methods may include application help functionality, wikis, formal training courses, and Audio/ Video presentations.

The Training Plan includes definition of the objectives of the training, the types of training required, the types of who will be trained, training formats by user audience, the schedule, costs, and a record of final approvals.

The Training Plan contains the following sections and topics:

- **Overview**
  - Introduction
  - Scope
- **Training**
  - Training Approach
  - Training and Environment Requirements
  - Training Courses
  - Training
  - Technical Support Training Course Topics
  - User Training Course Topics
  - Training Schedule
- **Approval and Signoff**

Training Plan
Project Name
Version

2.2 Training and Environment Requirements

Provide information about the training and environment requirements.

Topic	Description
Location	*Training will be performed onsite at the New York City headquarters location.*
Classroom	*The training room will contain the appropriate personal computer and presentation equipment for up to 8 students.*
Training Period	*During the last 2 weeks of Dec 2008*
Training Courses	*Two courses totaling 6 days.* o *Technical Support Training (4 days)* o *User Training (2 days).*
System Environment	*Training will be performed in a stable functional / integration environment. Personal computers will include technical and administrative modules / functions with remote access.*
Technical Student Skills	*Student personnel include developers and engineers.*
Lunch	*Students will be provided sufficient time to access the cafeteria for lunch.*

Note: The Trainer needs at least a half-day at the site prior to initial training to set-up the training environment.

2.3 Training Courses

2.3.1 Training

Provide information about the specific training:

- *Objective of the training course and the level of skill and understanding in the functions required.*
- *Topics or functions the course will review and cover.*
- *How long the training will last and the type of training, e.g., presentations along with demonstrations and exercises on a functioning system.*
- *Training prerequisites, e.g. familiar with Microsoft Client and Server technologies, an understanding of the corporate network topology and bandwidth.*

TRAINING REQUEST TEMPLATE

Training requirements for new or enhanced systems may be discovered on an ad hoc basis. In other words, after the software or system is deployed, various types of users and operations staff may identify functionality within the application that is not easily understood. In such cases they should initiate a training request process, which this training request template facilitates.

This template provides the ability to request training for a specific purpose, is relatively brief, and includes the following sections and data fields:

- **Training Request**
  - Name of requester
  - Position
  - Department
- **Training information**
  - Course Title
  - Training Objectives
  - Relevance to Roll/Responsibilities
  - Benefits to Staff Member
  - Review by Training Coordinator
- **Provider/Course Details**
  - Type Training (Internal versus External)
  - Turner/Provider
  - Course Date
  - Course Location
  - Duration (From/To dates)
- **Training costs and authorization.**

Training Request and Approval
Project Name
Version

1 Training Request

Name	Position	Department

2 Training Information

Information	Complete Entry
Course Title:	
Training Objectives:	
Relevance to Role/Responsibilities:	
Benefits to Staff Member:	
Review by Training Coordinator:	

3 Provider/Course Details

Type:	☐	Internal	☐	External
Trainer/Provider:				
Course Date:		Course Location:		
Duration	From:		To:	

SERVICE LEVEL AGREEMENT TEMPLATE (SLA)

SLA begins as the description of a services requirement or the description of an offering by a solution provider. The SLA is included in this requirements section as the service requirements exist and must be documented before a final agreement is executed.

The SLA formalizes an arrangement between a business organization or government agency and the service provider that will deliver the specified and agreed support services. The SLA documents the types and levels of support, the hours covered, the departments and numbers of people who are covered under the agreement, and contract costs and service delivery period.

The Service Level Agreement Template is used to identify the services that are required (and, if necessary, to identify services not included in the agreement), the identification of applications covered under the agreement, the required levels of effort, and the hours of operation. The SLA specifies required performance metrics that must be reported, and the frequency of reporting. And, the SLA stipulates how the agreement is amended, the conditions under which the agreement may be terminated, and how the agreement is renewed.

SLAs commonly include sections within the document to define services, performance measurements, problem management processes, support tiers, customer duties and obligations, warranties, and reliability and disaster recovery requirements.

The SLA must specify clear lines of demarcation governing respective client and supplier responsibilities. Contract enforcement (rewards and penalties) should be rigidly enforced, but most SLAs also leave room for annual visitation to allow changes based on new information.

Within the SLA template includes sections to detail requirements, terms and conditions over the life of the contract:

- Term of agreement
- Supported Organizations
- Approvals
- Key Contacts
- Organizational Dependencies
- Roles and Responsibilities
- Covered Applications
- Disaster Recovery Services
- Support Tier Services/ Response Times
- Days and Hours of Operation
- Maintenance Schedules
- Problem Resolution & Severity levels

Service Level Agreement Template
Project Name
Version

Note: Text displayed in blue italics is included to provide guidance to the author and should be deleted before publishing the document. In any table, select and delete any blue line text; then click Home→Styles and select "Table Text" to restore the cells to the default value.

1 Purpose

Purpose describes the intent of the document, which is to formalize an arrangement between your company and the client to deliver specific support services, at specific levels of support, and at an agreed-upon cost.

2 Scope of Agreement

This section provides the services included and excluded in the service level agreement.

2.1 Services Included in the SLA

Provide a list of services included in the service level agreement (specific service effort must be defined by your organization), e.g.,

Service	Description
Response to the transfer of trouble tickets	*Response to the transfer of trouble tickets for level 3 support, which includes the following:* • *Corrective Maintenance with root-cause analysis and emergency repair of any system operation.* • *Ticket status updates.*
Application monitoring	*Monitor production applications to assess application availability.*
Production application software enhancements	*Complete an enhancement to an existing production application, determine level of effort, e.g., less than five days.*
Transition of new or modified applications	*When a new or modified application is ready to be transitioned into support.*
Preventative maintenance	*Analyze and take steps to prevent potential problems. Criticality and priority determines activities that help determine corrective maintenance activity.*
Change management	*Manage new or changed processes, practices, or policies.*
Status reporting	*Prepare and communicate scheduled status reports and ensure management/team is aware of the support issues and risks faced.*
Knowledge management	*Record, store, and retrieve information to assist in problem resolution.*

System Design Phase
Overview

System Design is an overarching process to define the elements necessary to build a software-based system. The elements of software design include an architecture, the system's modules and underlying components, user and system-to-system interfaces, component operations, and data sources, inputs and outputs. The design process installs desired functionality within the software system to support identified business, user, and technical requirements.

The system design phase, within the traditional software development model, organizes work associated with defining an architecture and design for IT Software and systems under development. In simplest terms, a software architecture defines the structure of a proposed technical solution, while the software design provides the implementation details.

The architectural structure facilitates achievement of the non-functional requirements through the specification of necessary software and system components and their interfaces. The architecture may define the software or system's structure in multiple layers, such as Presentation/ User Interface, Business, Service, Data Access/ Middleware and Data Layers. The software architecture provides a big picture view of software and system requirements in context with the business and technology standards set by the organization.

Where the software architecture identifies the structure to support the non-functional requirements, the software design provides a more granular view of implementation details such as product features and functions, code, classes, components and software modules. More precisely, the designs describe the software modules, software components and classes necessary to implement the desired functionality, plus their respective data inputs and outputs, and behaviors. The design document also provides details on data sources, structures and designs, user interface designs, testing strategies, and analysis of issues that can limit or restrict the design. In short, the software design provides details on the construction of code, components and data that implement the identified business, user, and technical requirements.

The design phase starts with the functional and non-functional requirements as inputs to the architecture and design processes. Architecture assessments define the structure framework to build the software or system. Then, the developers define the design elements supporting each requirement or set of requirements.

Design elements describe the software functions and features in detail. Each design element provides details on functional diagrams, screen layouts, business rules, business process diagrams, and an entity-relationship diagram with a full data dictionary. Design elements describe the software in sufficient detail that the developer can build the software with minimal additional input.

The technical specialists begin to translate the requirements into detailed design solutions that create the systems, features, and functions that are necessary to achieve the business and functional requirements.

Numerous modeling methods and tools support the software and system engineering design process. Some of the most common approaches include:

- **Business Process Modeling Notation (BPMN)**
- **Data Flow Diagrams (DFDs)**
- **Functional Modeling**
- **Integrated Definition (IDEF) Modeling Languages**
- **Model-Based Systems Engineering (MBSE)**

- Language (SysML)
- UML (Unified Modeling Language)
- Process Modeling/ Flowcharts

IDEF, MBSE, SysML and UML are industry standard approaches that include multiple modeling techniques that span requirements analysis, functional modeling, data and information modeling, object-oriented modeling, ontology and semantics descriptions modeling, process and simulation modeling, and many other modeling techniques.

In the previous requirements definition phase, the focus of described work is on defining specific capabilities desired by and in support of the organization, and its customers and end-users. In the system design phase, the attention turns to defining systems and technical requirements.

This is also the phase where prototyping is likely to occur, as a prototype can serve as a bridge between requirements, design and development. A prototype can be as simple as a mockup of the proposed screen layouts, or more complex with the addition of business rules - instantiated as workflows and algorithms to manipulate data within the application, and with fields to capture and display data or information.

In the traditional software development model, on fairly large and complex projects, requirements gathering, analysis and documentation activities often takes 6 to 8 weeks to complete. Similarly, the Design Phase can also take 6 to 8 weeks to complete, typically with couple of weeks in overlap between the start of the Requirements and the Design Phase.

What this means is the first 10 to 12 weeks of the project, within the traditional software development model, are spent on requirements and design related activities before development starts. The project manager should plan to add in another 2 to 4 weeks for initial project planning. Therefore, development related activities may not start until 12 to 14 weeks into the project. But that does not mean prototyping cannot begin earlier - as part of the design phase related work activities.

Shortchanging requirements and design phase activities in the traditional development approach almost always lead to Scope Creep, as the project team will not have taken sufficient time to fully gather, analysis, document and vet their findings. Scope creep is where the project work expands beyond the approved budget and schedules – and is not a good thing to have happen.

Forms & Templates

SOFTWARE ARCHITECTURE PLAN (SAP) TEMPLATE

Sometimes referred to as a Software Architecture Document (SAD), the SAP – both very unfortunate acronyms, provide an approach to analyze structural requirements necessary to fulfil the non-functional/ technical requirements of the product. Non-functional requirements are also known as quality of service (QoS) requirements, service-level requirements and technical requirements.

Non-functional requirements specify operational capabilities of a system in the form of qualitative requirements. Examples of qualitative assessments includes accessibility, adaptability, availability, extensibility, interoperability, maintainability, operability, portability, reliability, reusability, scalability, and supportability. In other words, these specifications specify the "*abilities*" the software or system that the architecture must accommodate.

Of course, a non-functional criterion does not have to include "abilities" in its descriptive name. There are many examples of non-functional criterion without "abilities" in the name, such as backup, disaster recovery, failover, fault tolerance, open source, performance, and others.

Software architectures often employ *views* or *layers* to analyze and develop system structures. The view model describes the architecture from different viewpoints most relevant to the issues addressed by the application, such as logical, process, physical, development, operational, information, functional, engineering and technology. An architecture layer model breaks the structure of the application into separate and independent tiers to isolate related functionality, such as presentation, business, persistence, access, and data source layers.

When completed, the SAP/ SAD provides a comprehensive architectural overview of the system and implements a number of different architectural views and/ or layers to depict important structural aspects of the system. The SAP/ SAD assessments also capture and communicate the significant architectural decisions made by the architects in support of the systems design. The Software Architecture Plan contains the following topics:

- **Scope**
  - Logical view
- **Definitions, acronyms and abbreviations**
  - Overview of design model
- **References**
  - Process view
- **Architectural representation**
  - Deployment view
- **Architectural goals and constraints**
  - Implementation view
- **Architecturally significant design packages**
  - Data view
- **Overview of document**
  - Size and performance
- **Use-Case view**
  - Quality

YOUR COMPANY LOGO

Software Architecture Plan
Project Name
Version

This document provides a comprehensive architectural overview of the system, using a number of different architectural views to depict different aspects of the system. It is intended to capture and convey the significant architectural decisions which have been made on the system.

Note: Text displayed in blue italics is included to provide guidance to the author and should be deleted before publishing the document. In any table, select and delete any blue line text; then click Home→Styles and select "Table Text" to restore the cells to the default value.

1 Purpose

The purpose defines the role or purpose of the Software Architecture Document in the overall project documentation, and briefly describes the structure of the document. The specific audience for the document is identified, with an indication of how they are expected to use the document.

1.1 Scope

This is a brief description of what the Software Architecture Document applies to; what is affected or influenced by this document.

1.2 Definitions, Acronyms, and Abbreviations

Provides the definitions of all terms, acronyms, and abbreviations required to properly interpret the Software Architecture Document. This information may be provided by reference to the project's Glossary.

1.3 References

Provides a complete list of all documents referenced elsewhere in the Software Architecture Document. Identify each document by title, report number (if applicable), date, and publishing organization. Specify the sources from which the references can be obtained. This information may be provided by reference to an appendix or to another document.

1.4 Overview

Describes what the rest of the Software Architecture Document contains and explains how the Software Architecture Document is organized.

Confidential - ©2015 Documentation Consultants (www.SDLCforms.com) Document: 3200 Page 4 of 7

www.SDLCforms.com

SYSTEMS REQUIREMENTS SPECIFICATIONS (SRS) TEMPLATE

The SRS is a technical design document that specifies the information necessary to construct a software-based system that implements the capabilities defined within the functional requirements. Functional requirements define system, module and component level behaviors and their inputs and outputs.

The SRS provides more details on how the application implements business and user requirements as components, features and functions, defines the physical characteristics of the user and system interfaces, and documents other critical design considerations. Typical design considerations include target hardware, network, and software operating environments, security components, compliance with regulatory mandates, disaster recovery capabilities, manufacturing specifications, and data specifications.

An SRS template may contain the following sections and underlying design specifications, and other specifications as required:

- **Introduction**
    - Purpose
    - Scope
    - Roles and Responsibilities
- **System Requirements Specifications**
    - Product Functional Specifications
        - Functional System Specifications
        - Software Specifications
        - Hardware Specifications
    - User Characteristics / Ergonomic Specifications
        - System - User Ergonomics
    - Operating Environment Specs
        - Computer Systems Environmental Specifications
    - Security Specifications
        - Physical Hardware Security Specifications
        - System Security Specifications
        - Network Security Specifications
    - Regulatory Specifications
    - Disaster Recovery Specifications
        - Disaster Recovery Specifications
        - Business Resumption Specifications
    - Manufacturer Specifications
    - Data Specifications
    - Network Impact Specifications.

System Requirements Specifications
Project Name
Version

1.3 Roles and Responsibilities

The following table provides the roles and responsibilities for this project.

Examples:

Roles	Responsibilities
Author	• *Prepares the document and distributes for review.* • *Signs the document attesting to its contents.* • *Ensures that all information contained within the document is correct.* • *Participates in the Analysis phase of the SDLC.* • *Updates document, as necessary, throughout the SDLC process.*
Client	• *Reviews and approves this document.* • *Participates in expanding high level requirements into system requirements specifications.* • *Participates in design of business process definitions.* • *Participates in training and data conversion activities.* • *Participates in maintaining the system following production release.*
Project Manager	• *Reviews and approves this document.* • *Ensures that all information contained within the document is correct.* • *Participates in all phases of the SDLC.* • *Ensures change control procedures are adhered to.* • *Tracks and communicates problems found during design, construction, configuration, implementation, and maintenance of the system.*
Technical Rep	• *Provides expertise in the development of all technical documentation.* • *Participates, reviews, and approves this document.*
QA Manager Rep	• *Ensures that policies and procedures are followed.* • *Participates, reviews, and approves system requirement specifications.* • *Coordinates and/or performs the Quality Assurance activities.* • *Participates in all SDLC phases.* • *Serves as a subject matter expert for related requirements.* • *Monitors that control has been maintained after release for production.*
Change Control Manager	• *Ensures the Change Control process is followed.*
Project Team	• *Reviews and approves this document.*

ANALYSIS AND DESIGN DOCUMENT

The previously described SAP and SRS documents provide tools to document architecture and design decisions. In contrast, the Analysis and Design Document (ADD) implements a process to analyze and define architecture and design requirements. Specifically, the ADD provides detailed guidance on how to perform an analysis and design of a system, including analysis of current and future software architecture processes, interfaces, data flows, infrastructures, components, integration/ middleware components, and security.

Systems analysis takes a systemic view to analyzing complex processes or operations in order to improve their efficiencies. From the perspective of building and refining software applications and IT-based systems, systems analysis evaluates the functionality, effectiveness, and efficiencies of the system as a whole. E.g., as a set of interoperating components.

The ultimate goal of systems analysis is to ensure a product is useful and efficient. But the process of systems analysis also helps identify those areas within both the supported business process and the supporting business systems or application that can be improved.

The Analysis and Design Document contains the following sections and subtopics:

- **Overview**
  - Purpose and Scope
  - Referenced Documentation
  - Design Assumptions and Support Considerations
- **System Overview**
  - Summary of Changes
- **Analysis**
  - Business Impact
  - Application Impact
  - Current Application Behavior
  - Current Software Architecture
    - Current Process Model
    - Current User Interface
    - Current Data Model
  - New Software Architecture
    - New Process Model
    - New User Interface
    - New Data Model
    - Procedures
    - Data Changes
- **Security and Audit**
- **Architecture Design**
  - Interface Design
- **Design, Development, and Integration**
  - Application Layer Information
  - Implementation and Software Execution
- **Infrastructure**
- **Future improvements**
- **Approvals**

Analysis and Design Document
Project Name
Version

Note: Text displayed in blue italics is included to provide guidance to the author and should be deleted before publishing the document. In any table, select and delete any blue line text; then click Home→Styles and select "Table Text" to restore the cells to the default value.

1 Overview

Briefly state the system, infrastructure, and software to which this document applies.

- *Summarize the history of system development, operations, and maintenance.*
- *Identify the project sponsor, user, developer, and support groups.*
- *Identify current and planned operating sites.*
- *High-level overview and technical summary of implementing the business requirements.*
- *List other relevant documents.*

1.1 Purpose and Scope

Provide information about the purpose and scope of this analysis and design document. Indicate what the design is intended to do and where it fits into the overall system architecture. Indicate the information that will be included and what is out of scope for this project, e.g., it may include information about development and infrastructure but not production support.

1.2 Referenced Documentation

Provide information about all documentation referenced in this document, e.g., number, title, version, and date.

Document Number	Version	Date	Document Name

APPLICATION DEVELOPMENT PROJECT CHECKLIST

The Application Development Project Checklist was originally developed to support the linear sequential planning process within the traditional software development model. Nevertheless, the checklist still has value in nontraditional and agile-based software development approaches by identifying the types of activities within the standard classifications of work.

The Application Development Project Checklist identifies 50+ tasks across typical work classifications, such as by definition, analysis, design, development, testing, implementation, and project close. The checklist includes fields to record who is responsible for each identified task, name of person who completed the work, completion dates, and any relevant comments.

Implemented as a project management planning tool, the checklist helps project and development managers ensure the development team is meeting project timelines and deliverables. First and foremost, the tool provides a quick view of the tasks and functions within the respective work stages. Whether used formally or informally, the checklist helps provide more rigor in identifying and following through with the completion of product life cycle development tasks.

The Application Development Project Checklist contains the following sections and checklists:

- **Purpose**
- **Product or System Information**
  - Product or System Overview
  - Reason and Description
  - Assumptions, Dependencies, and Constraints
  - Stakeholders
  - Risks
  - Implementation Dates
- **Application development checklist:**
  - Project definition
  - Analysis phase
  - Design phase
  - Development phase
  - Testing phase
  - Implementation phase
  - Evaluation phase.

Application Development Project List
Project Name
Version

3 Application Development Checklist

Fill in the following table with the appropriate information.

Description	Responsibility	Completed by	Date	Comments
Project Definition				
Describe Problem or Opportunity				
Describe Business Objective and Scope				
Create Business Case (if necessary)				
Identify Related Projects				
Identify Project Team				
Describe High Level Time Line				
Define Project Costs / Budget				
Assess Risks				
List Assumptions				
Sign Project Charter				
Develop Schedule for Analysis Phase				
Analysis Phase				
Create Project Directory on Server / LAN				
Create Issues Document				
Document Business Requirements				
Document Systems Requirements				
Modify Project Charter (if needed)				
Create Schedule for Design Phase				
Design Phase				
Identify Nature of Project				

TECHNICAL REQUIREMENTS DOCUMENT (TRD)

It's sometimes useful to capture technical/ non-functional requirements as a separate activity apart from developing the software or system architecture. This keeps the focus on identification of technical requirements and related structural components before trying to figure out how to structure the application to support the technical requirements.

The TRD summarizes the technical requirements for a software application or system and its related infrastructure requirements. Specifically, the TRD supports documentation of technical capabilities the system must support, such as the total complement of activities, hardware, software, test, integration, functions, features, capabilities, operations, performance-related issues, reliability issues, and availability issues.

A TRD contains technical implementation details, such as:

- Descriptions of the software's purpose and functionality
- Documentation of non-functional requirements (e.g., MTBF, response time, availability, portability, maintainability, recovery speed, and other important criteria.)
- Use cases descriptions of user and systems interactions
- Identifies hardware, network, security and system failover requirements
- Performance engineering requirements
- Quality standards
- Design constraints

The Technical Requirements Document contains the following sections and topics:

- Overview
  - Purpose and Scope
  - Referenced Documentation
  - Requirement Assumptions and Support Considerations
- Specific Technical Requirements
  - Technical Requirements
    - System Requirements
    - Network Requirements
    - Database Requirements
    - User Interface Requirements
    - System Interfaces Requirements
    - Security Requirements.

YOUR COMPANY LOGO

Technical Requirements Document
Project Name
Version

2 Specific Technical Requirements

This section contains all the requirements in sufficient detail that can be used to develop a system design and allow testers to test the system.

2.1 Technical Requirements

This area focuses on the technical requirements that are required or needed within the system.

- *Mandatory - "Must Have" Represents the core functionality and must be included in the solution.*
- *Desirable - "Nice to have" Requirements " Once all mandatory requirements are met and sufficient resources are available.*

2.1.1 System Requirements

Provide the overall system requirements.

2.1.2 Network Requirements

Provide the network requirements.

2.1.3 Database Requirements

Provide the database requirements.

2.1.4 User Interface Requirements

Provide the user interface requirements.

2.1.5 System Interface Requirements

Provide the system interface requirements.

2.1.6 Security Requirements

Provide the security requirements.

DATABASE DESIGN DOCUMENT

The Database Design Document maps a logical data model to the target database management system while considering the system's performance requirements. Specifically, the database design document is a tool that helps convert logical or conceptual data constructs to physical storage constructs (e.g., tables, files) of the target database management system (DBMS).

The Database Design Document contains the following sections and related topics:

- **Document objectives**
- **Intended audience**
- **Key personnel**
- **Data owners**
- **Assumptions, constraints and dependencies**
- **System Overview**
  - Database management system configuration
  - Database software utilities
  - Support software
- **Database-Wide design decisions**
  - Interfaces
  - Key factors influencing design
  - Behavior
  - DBMS platform
  - Security and availability
  - Distribution
  - Backup and restore operations
  - Maintenance
- **Database administrative functions**
  - Responsibilities
  - Database identification
  - Applications / systems using the database
  - Relationships to other databases
  - Schema information
  - Physical design
  - Physical structure
  - Entity mapping
  - Mapping rules
  - Operational implications
  - Data transfer requirements
  - Data formats
  - Business rules
  - Storage
  - Backup and recovery
- **Architecture**
  - Hardware architecture
  - Software architecture
  - Datastores
- **Detailed database design**
  - Data software objects and resultant data structures
  - Database management system files
  - Reporting requirements.

Database Design Document
Project Name
Version

Note: Text displayed in blue italics is included to provide guidance to the author and should be deleted before publishing the document. In any table, select and delete any blue line text; then click Home→Styles and select "Table Text" to restore the cells to the default value.

1 Purpose

The Database Design Document maps the logical data model to the target database management system with consideration to the system's performance requirements. The Database Design converts logical or conceptual data constructs to physical storage constructs (e.g., tables, files) of the target Database Management System (DBMS).

1.1 Document Objectives

The Database Design Document has the following objectives:

- To describe the design of a database, that is, a collection of related data stored in one or more computerized files that can be accessed by users or computer developers via a DBMS.
- To serve as a basis for implementing the database and related software units. It provides the acquirer visibility into the design and provides information necessary for software development.

1.2 Intended Audience

This document is intended for the following audiences:

- Technical reviewers, who must evaluate the quality of this document.
- Developers including:
  - Architects, whose overall architecture design must meet the requirements specified in this document.
  - Designers, whose design must meet the requirements specified in this document.
  - Developers, whose software must implement the requirements specified in this document.
  - Quality Assurance personnel, whose test cases must validate the requirements specified in this document.

1.3 Acronyms and Abbreviations

Acronym / Abbreviation	Meaning
RDMS	Relational Database Management System
DBA	Database Administrator

WEBSITE PLANNING CHECKLIST

The Website Planning Checklist is a tool to review design considerations when designing and developing a new website. Whether used formally or informally, the checklist helps ensure important website design decision criterion are not forgotten before starting development. Taking the time to analyze website design considerations will reduce time-consuming and costly re-work once development begins.

The Website Planning Checklist helps keep the design focus on achieving the goals and objectives of the web site. It helps the developers develop the user capabilities and services required, and not on working on building services that are not value-added.

The Website Planning Checklist contains the following website design sections and design considerations:

- **Purpose and Introduction**
- **Website Planning Checklist:**
  - Purpose, Goals and Objectives
  - Audience Analysis
  - Audience Profile Checklist
  - Strategy - Competitor Analysis
  - Strategy - Determine your approach
  - Site content
  - Advertising and maintenance
  - Organizing information
  - Site structure
  - Navigation
  - Text
  - Visual Design and layout
  - Images
  - User interface design techniques
  - Testing

Website Planning Checklist
Project Name
Version

2 Website Planning Checklist

Your first planning question should be "What is the purpose of the website?

- What do you want to do?
- What do you want to sell or accomplish on the web.

Use the checklist as a guide, which will influence your user interface design.

Topic	Comments
Purpose	
What do you want to accomplish?	
• *Provide a clear statement of goals.* • *More customers.* • *New geographical markets.* • *Increased sales.* • *Greater customer loyalty.* • *Offer customers a better way to buy.* • *Something that's not possible by traditional means.*	
Audience Analysis	
Which user groups?	
Business to Business (B2B) • *Large supermarket chains* • *Small shops* • *Catering companies* • *Suppliers.*	
Business to Customer (B2C) • *What kind of audience?* ○ *Rich, selective, specialist* ○ *Lower class, cheap with bulk* ○ *Young, modern market.*	
Which of these users have web access and are likely to use it?	
Obtain feedback from potential users.	

USER INTERFACE (UI) DESIGN TEMPLATE

The User Interface Design Template provides a tool to document user requirements that are necessary to define screen designs, or redesign existing display screens, prior to the start of any programming activities. The discipline of user interface design applies to both web page designs and traditional software application display screens.

The UI document implements a method and a tool to define the screen details (names, screen description, major purpose/function, etc.), tab design, definition of each field, action/command buttons, any special rules, actions, formats, color schemes, popup and error messages, and navigation rules.

The User Interface Design Template contains the following sections:

- The application or product name
- The product or system overview
- The reason for the redesign or new product development
- The screen name and processing to be performed by the screen
- Any assumptions, dependencies or constraints on the design
- Risks
- List of stakeholders
- Security restrictions.

The following parameters are identified for each screen design:

Field Type	Field type and name
Data Type	Data types
Length	Associated functions and processes
Required Fields	Define required vs. optional fields
Calculations	Define any calculations or formulas
Display Only Fields	Define the databases and tables to be utilized
Dropdown/List Boxes	Define all dropdown, list or combo boxes
Field navigation	Field tab sequence
Default Data	Identify special or unique processes, functions, fields or actions
Font name and style	Font name and regular, bold, bold italic or italic
Font color and size	Define color code and point size
Rules	Any special rules governing development
Buttons	Define all command or action buttons
Pop-up messages	Define action and content of popup messages
Formats	Date and time formats
Events	Define next event to occur

User Interface Design Template
Project Name
Version

3 Screen Details

Screen Information:

Topic	Information
Main Screen Name	
New or Existing Screen	
Screen Description	
Major Function	*Indicate the major function that the screen is used for, e.g., find feature, generate reports, Other (explain).*
Initial Access activity	*Indicate the activity that must be performed first before you can navigate to the screen, if any, e.g., Find function must be performed first before account maintenance.*
Security access restrictions	*Indicate any special access restrictions, e.g., Finance department only, managers and above, etc.*

Tabs on the screen that lead to other sub-screens:

Tab Names	Information

Confidential – ©2015 Documentation Consultants (www.SDLCforms.com) Document: 4300 Page 6 of 13

www.SDLCforms.com

REPORT DESIGN TEMPLATE

Large and complex software applications, especially those used to support business processes and functions, often maintain tremendous amounts of data within their databases. Managers and users often need access to this data in report type formats. It may also be necessary to implement business intelligence and other analytics against the data, stored within the application's databases, before generating the reports. The report design template helps document identified management and user requirements on data analytics and the reports they expect the application to produce.

The report design document implements a method as a tool to design and develop reports and supporting data analytics. Within the report design template, the analyst notes the report name, the reason the report is required, the type of data required within the report, data analytics, and the format of the output format. The report design template also records the report's audience, who can run the report, the frequency of report generation, report input parameters (e.g., parameters, input data type, and text input), report fields (e.g., field name, description, database or table source, calculated field, the formulas and or functions), group field information, sort order, page header and footer contents.

The Report Design Template typically contains the following sections and subtopics:

- **Purpose**
- **Product or System Information**
  - Product or system overview
  - Report Design or Development Reason and Description
  - Assumptions, Dependencies, and Constraints
  - Stakeholders
  - Risks
  - Implementation Dates
- **Report Details**
  - Report Name
  - Report Description
  - Intended Audience
  - Security/Access
  - Frequency
  - Who can run this report?
  - Where is the report run?
  - Report Input Parameters and Selections
  - Report fields Group Field Information
  - Sort Order Information
  - Page Header and Footer Information
  - Report Header and Footer Information
  - Report Sample (Graphic example of report's layout)

YOUR COMPANY LOGO

Report Design Template
Project Name
Version

Selection Input Parameters	Input Data Type (e.g., number, text, date)	Text Input

Report Fields:

List each field that will be listed on the report.

Field Names on the Report	Description	Indicate Database or Table data is from (if known)	Is Field Result based on a calculation (Y/N)	If yes, describe the formula / calculation and/or function it comes from

CODE REVIEW CHECKLIST

Code reviews are necessary in both the traditional and agile-based software development projects. They provide a quality check to ensure the code is readable, follows the organizations coding standards, is not duplicated elsewhere, is testable and tested, and efficient.

Within a large organization, a reusable code checklist template can be quite large. For example, the Code Review Checklist offered by SDLCforms contains nearly 140 checklist items. Each development team may not need to use all the checklist items maintained within the organization's master list, and they should assess the code checklist items to extract those that are germane for their architecture, designs, code and testing requirements related to their product.

A Code Review Checklist provides a company standard guideline for checking code and should include fields to indicate the pass/ fail of each evaluated checklist item, and fields to record comments for each code test item.

The code checklist items within the SDLCforms template spans each of the following categories and related sections:

- [Code] Structure
- Documentation
- Variables
- Style
- Architecture
- Arithmetic operations
- Loops and branches
- Defensive Program
- Maintainability
- Requirements and Functionality
- System and Library Calls
- Reusability
- Robustness
- Security
- Control Structures
- Resource Leaks
- Error Handling
- Timing
- Validation and Test
- Hardware

Code Review Checklist
Project Name
Version

The Code Review Checklist provides a company guideline for checking code including pass/fail parameters and recording any comments when the test fails.

During a project, this document is used by team members as follows:

1. During project planning, it is utilized as a reminder for how much review time should be allocated during the project for the software being developed
2. During design and development (coding) portion of the project, the checklists are used to conduct code reviews.

The list of test items is representative and should be modified prior to use to reflect your development environment and standards.

Generic Checklist for Code Reviews

Structure

	Description of Item	Pass	Fail	Comments
☐	Does the code completely and correctly implement the design?	☐	☐	
☐	Does the code conform to any pertinent coding standards?	☐	☐	
☐	Is the code well-structured, consistent in style, and consistently formatted?	☐	☐	
☐	Are there any uncalled or unneeded procedures or any unreachable code?	☐	☐	
☐	Are there any leftover stubs or test routines in the code?	☐	☐	
☐	Can any code be replaced by calls to external reusable components or library functions?	☐	☐	
☐	Are there any blocks of repeated code that could be condensed into a single procedure?	☐	☐	
☐	Is storage use efficient?	☐	☐	
☐	Are symbolics used rather than "magic number" constants or string constants?	☐	☐	
☐	Are any modules excessively complex and should be restructured or split into multiple routines?	☐	☐	

Documentation

	Description of Item	Pass	Fail	Comments
☐	Is the code clearly and adequately documented	☐	☐	

CONVERSION PLAN

The software industry is ever-changing in software and systems quickly become legacy systems. At some point those legacy systems become unsupportable and unsustainable from a cost, skills, and resource perspective. However, the business purpose of those applications lives on. At those times the IT organization must assess the viability of building new systems to replace the legacy systems, or to replace old technology with new. The conversion plan provides a method and a tool to plan and implement system the data conversion strategies.

The Conversion Plan provides critical information on the following topics:

- **Identifies impacted organizations, processes, and systems**
- **Communicates the strategy for a successful conversion**
- **Provides a roadmap for performing the conversion**
- **Outlines the total scope of work**
- **Documents the conversion schedule**
- **Identifies important pre and post conversion plans and activities**

The Conversion Plan contains the following topics:

- Introduction
  - Purpose and Scope
  - Referenced Documentation
- Conversion Information
  - System Overview
  - System Conversion Overview
  - Conversion Description
  - Types of Conversion
    - Conversion Strategy
    - Conversion Risk Factors
  - Conversion Tasks
    - Conversion Planning
    - Pre-Conversion Plans
    - Major Tasks and Procedures
  - Conversion Schedule
  - Security
- Conversion Support
  - Hardware
  - Software
  - Facilities
  - Materials
  - Personnel

YOUR COMPANY LOGO

Conversion Plan
Project Name
Version

- *Conversion of different versions of a software system, e.g., a database management system (DBMS), operating system, or local area network (LAN) management system.*
- *Conversion from one computer language to another or from one software system to another.*

2.4.1 Conversion Strategy

Describe the strategies to convert system hardware, software, and data.

- *Hardware Conversion Strategy: Includes the new hardware environment.*
- *Software Conversion Strategy. Includes the software conversion strategy to be used.*
- *Data Conversion Strategy. Includes the strategy for data conversion, data quality assurance, and the data conversion controls.*
- *Data Conversion Approach. Includes data preparation requirements and the data that must be available for the system conversion. Provide a detailed description of the data handling, conversion, and loading procedures if data will be transported from the original system.*
- *Interfaces. Includes the new or affected interfaces. For a hardware platform conversion, e.g., mainframe to client / server, provide information about affected interfaces and their revisions.*
- *Data Quality Assurance and Control. Includes assurance for data quality before and after all data conversions, e.g., approach to data scrubbing and quality assessment of data before they are moved to the new or converted system.*

Strategy Topic	Description
Hardware	
Software	
Data Conversion	
Data Conversion Approach	
Interfaces	
Data Quality Assurance and Control	

2.4.2 Conversion Risk Factors

Describe the major risk factors in the conversion and strategies for their control or reduction, e.g.,

- *Technical performance of the converted system*
- *Conversion schedule*
- *Costs*

Testing Phase
Overview

Software system testing helps ensure the product ultimately achieves the goals and objectives that justified the development investments, and that the products implement the functional and nonfunctional capabilities identified by customers and users. Without software testing, it's impossible to know for sure how well a product meets the requirements and their acceptance criteria. In effect, software and stems testing is part of the organization's quality assurance processes.

Whether the organization uses traditional or agile-based software development practices, testing is product lifecycle activity. In other words, activities to support product testing began at the earliest phases of work and continue on long after the product is deployed.

As a process, testing typically begins with the development of acceptance criteria for each requirement. Developers then create test scripts and test scenarios for each requirement that, when executed against the code and software modules, indicate conformance with the requirements and their acceptance criteria. Initial testing activities commence in parallel with development activities, starting with unit and integration testing. System and performance related tests follow. Test results are analyzed to ensure the code, components or modules, and products implement the desired capabilities in conformance with their acceptance criteria.

- There are many types of tests performed across the product development lifecycle. Some of these tests include:
- Unit testing - developer tests performed on each block of code they write
- Integration testing - to ensure the new code doesn't break existing code or functionality
- System testing - evaluates the system implements all requirements as an integrated solution
- End-to-end testing - to ensure a business process or workflow is successfully implemented
- Acceptance testing - performed by users to ensure the system works and meets their requirements
- Usability testing - also performed by users, but with the goal to test the intuitiveness of the product
- Load testing - stresses the product to see if it breaks or performs poorly under increasing demands
- Regression testing - to ensure changes in code or components do not break original functionality
- Backup and recovery testing - to ensure timely recovery of data and the system if it fails
- Migration testing - to ensure data and applications can move to a new operating environment

- **Functional testing – to ensure that the software conforms with all identified requirements**
- **Performance testing – tests system speed, behaviors, and reliability under normal usage parameters**
- **Stress testing – automated tests of non-functional requirements under extreme loads**

Product testing normally starts with the plan. Not all products require the same types for level testing. The larger and more complex an application more testing it will require. The test plan provides the development team's assessments of the type of tasks necessary over the course of their work.

Many people are involved in testing. The developers typically test the source code as they build and integrate the code within their source code repository. End-users test and validate conformance of the software in context with the defined test scenarios that typify use within their business operations. End-users will evaluate the functionality of the user interfaces, screen navigations, data fields, data flows, and reports generated by the application. On larger software systems development projects, technical testers evaluate the underlying hardware, software, network, database, workflow, and security components and their conformance to architectural, design and performance requirements.

Independent Software or Systems Quality Assurance (SQA) tests confirm the development team has followed the organization's quality processes and procedures, and that the solution meets established quality metrics. Another independent group might conduct independent verification and validation (IV&V) Tests to verify the solution meets requirements and validate that the system meets the customer's needs.

When defects and bugs are found, they must be documented within the development team's defect/ bug logs and reports. Errors in the code that implements functionality that is not in conformance with specifications. In contrast, defects are identified failures to implement the capabilities not previously identified. That definition for defects may seem strange, but it's not unusual for customers and users to not identify product requirements or opportunities for improvement until they have a chance to use the product.

Defects and bugs that are uncovered during testing must be documented and prioritized. All critical bugs should be fixed prior to going into production. Defects are prioritized and implemented in future development iterations.

This section of the book identifies a number of methods and tools to create various software, system, and website testing plans, test scenarios, bug and defect reports and logs, and quality assurance checklists.

Forms & Templates

DOCUMENTATION QA CHECKLIST TEMPLATE

Under the traditional software development model, documentation is a big part of how organizations manage and report progress and conformance of work against the overall SDLC process. Because documentation is such a big part of the traditional model, it's not unusual to see a software quality assurance activity created just to check the accuracy and completeness of the documentation. The documentation QA checklist helps guide the SQA staff through their assessment of each document.

The primary objective of the Documentation Quality Assurance Checklist is to ensure a thorough documentation review prior to delivery and deployment of the production ready release of a product. The SQA review confirms the product documentation is technically correct and complete, uses a consistent format, is written at the appropriate audience level, and is free or spelling and grammatical errors.

- **The Quality Assurance (QA) Checklist provides guidance through each of the following document review requirements:**
  - QA Topic:
  - Document Properties
  - Track Changes
  - Title Page
  - Table of Contents
  - Headers and Footers
  - Heading and Section Titles
  - Spelling and Grammar Check
  - Document content, format, and style
  - Acronyms
  - Appendices
  - Contact Information
  - Cross-References
  - Footnotes
  - Graphics
  - Hyperlinks
  - Index
  - Page and Section Breaks
  - Page Numbers
  - Process and Workflow Charts
  - Special Characters
  - Table of Figures
  - Tables and Charts
  - Terminology.

Documentation QA Checklist
Project Name
Version

Note: Text displayed in blue italics is included to provide guidance to the author and should be deleted before publishing the document. In any table, select and delete any blue line text; then click Home→Styles and select "Table Text" to restore the cells to the default value.

Documentation Quality Assurance (QA) Checklist

All documentation should be reviewed for quality assurance prior to delivery and implementation. It is imperative that the documentation is technically correct and complete, uses a consistent format, written at the appropriate audience level, and is free of spelling and grammar errors.

QA Topic	Verified (yes/no)	Comments
Document Properties - *Click 'File' and 'Properties')-contains title, subject, author, and company.*		
Track Changes - *Special side bar characters on pages to denote changes, if applicable). All tracked changes are finalized and special side bar indicators, underlines, colors, etc. must not be displayed unless requested by the client.*		
Title Page – *Title, date, logo, names (company, preparer, and recipient names), confidentiality statement, and proprietary marking).*		
Table of Contents – *Numbered and aligned correctly, no bookmark errors, displays appropriate number of heading levels, and page numbers).*		
Headers and Footers – *Logo, date, page number, confidentiality statement, and proprietary markings on all pages).*		
Heading and Section Titles - *Correctly titled, numbered, and aligned with correct size and font).*		
Spell and Grammar Check - *(Note: not all recommended changes should be implemented).*		

TEST SCENARIOS TEMPLATE (A.K.A. BUILDING TEST SCENARIOS)

Before testing code it's useful to develop an approach testing the implemented requirements. There are three traditional approaches:

1. Test Scenarios – describes what end-to-end functionality is to be tested - decomposes into test cases

2. Test Cases – descriptions of how to execute a single test

3. Test Scripts – step by step set of instructions to execute a test, often used to automate testing

Testing scenarios implement a potential use of the software application or system through a complex problem or end-to-end execution of a business process or workflow. During acceptance testing, test scenarios guide the users through a typical business application, making it easier to evaluate the effectiveness of the application to support their business and user needs. In fact, it's not unusual for prospective users to be involved in defining the test scenarios.

Test Scenarios Template shown on the next page includes data fields to record the following information:

- **Test Scenario**
  - Version #
  - Build #
  - Test Scenario ID
  - Process ID
  - Environment
  - Machine tested on
  - Test scenario description
  - Objective of the scenario
  - Assumptions and constraints
  - Identification of any test files or test data
  - Author
  - The last date modified
  - Reviewer's name and date
  - Executor's name and date

- **Test steps**
  - Step #
  - Descriptions
  - Expected Results
  - Actual Results
  - Pass/ Fail

-

Building Test Scenarios
Project Name
Version

Note: Text displayed in blue italics is included to provide guidance to the author and should be deleted before publishing the document. In any table, select and delete any blue line text; then click Home→Styles and select "Table Text" to restore the cells to the default value.

1 PURPOSE

A test scenario is a hypothetical story, used to assist a person think through a complex problem or system.

The development of test scenarios helps the design team understand the motivations of users and the functions that will compliment those motivations. This is valuable for all tests, but especially important for scenarios because they are complex.

Scenarios are also useful to connect to software requirements, especially requirements modeled with use cases, and to reveal requirements-related controversies. Test scenarios also check if all of the business flows are working as expected.

2 DIFFERENCE BETWEEN A TEST CASE AND TEST SCENARIO

A test case is "How to be Tested," while a Test Scenario is "What is to be Tested."

The difference between a test case and a test scenario is explained below:

- A test case is a set of conditions or variables under which a tester will determine whether an application, software system or one of its features is working as it was originally established for it to do.
- A test scenario tells a story about how a user will interact with an application. A test scenario makes sure that end to end functionality of an application under test is working as expected. It also checks if the all business flows are working as expected. In scenario testing, the tester needs to put his/her mind in the end users mind to check and perform the action as how they are using the application under test. In scenario testing, the preparation of scenarios would be the most important part, in which the scenario tester needs to consult or accept help from the client, stakeholder or developers.

TEST PLAN TEMPLATE

A test plan provides details on the scope of testing and types of testing required to test a software application or systems under development. The test plan identifies the testing schedules, necessary resources – human, hardware and software, and approach to the testing.

The test plan must identify the features or workflow to be tested. In that sense, it's a living document that must be updated reflecting the impact of changes to the functional, non-function and design requirements. Testing subplans should be defined for each type of testing implemented on the project. So, for example, a regression test requirement is described within a regression test subplan. Likewise, for system, end-to-end, load, performance, stress and other testing types identified for the product.

The Test Plan shown on the next page includes the following sections and subtopics:

- **Introduction**
    - Features to be tested
    - Features not requiring testing
    - Testing approach.
    - Identifying and justifying tests
        - Unit test
        - Integration test
        - User acceptance test (UAT)
        - Load test
        - Operational readiness test
        - Beta testing
    - Measuring the extent of testing
        - Entrance criteria
        - Exit criteria
- **Dependencies, assumptions and constraints**
    - Dependencies
    - Assumptions
    - Constraints
- **Notification and escalation procedures**
- **Measures and metrics**
    - Test preparation
    - Test execution and progress
    - Bug analysis
- **Testing tasks**
- **Suspension criteria and resumption of testing**
- **Status and issues**
- **Approvals**

YOUR COMPANY LOGO

Test Plan
Project Name
Version

For each testing phase, a detailed test plan shall be developed that identifies the testing requirements specific to that phase.

Specific items to be identified in each test plan shall include:

- Test Items
- Test Execution Procedures
- Test Deliverables
- Test Data Management
- Test Schedule
- Test Environment

1.4 Identifying and Justifying Tests

Describe how tests will be identified and considered for inclusion in the test effort covered by this strategy. Include the sections below that will be in scope for testing your specific project. Unit Test, Integration Test, and User Acceptance are required

1.4.1 Unit Test

Unit testing is the initial testing of new and/or changed code in the system. The purpose of unit testing is to allow the developer to confirm the functionality provided by a single unit or component of code. Additionally, wherein one component cannot function without interacting with another component, the test shall include limited interactions.

Unit testing shall consist of the following:

- **Static testing –** Conducting "walkthroughs" and reviews of the design and coded components.
- **Basic path testing –** Executing path testing based on normal flow.
- **Condition/multi-condition testing –** Executing path testing based on decision points.
- **Data flow testing –** Examining the assignment and use of variables in a program.
- **Loop testing –** Checking the validity of loop constructs.
- **Error testing –** Executing unexpected error conditions.

1.4.2 Integration Test

Integration testing confirms that each piece of the application interacts as designed and that all functionality is working. Integration testing includes interactions between all layers of an application, including interfaces to other applications, as a complete end-to-end test of the functionality.

Integration testing shall consist of the following:

- Verifying links between internal application components.
- Focusing on complete end-to-end processing of programs, threads, and transactions.
- Boundary value analysis (testing modules by supplying input values within, at, and beyond the specified boundaries).
- Cause-effect testing (supplying input values to cause all possible output values to occur).

SYSTEM QUALITY ASSURANCE (SQA) CHECKLIST

Software or system quality assurance (SQA) is an approach to check whether quality objectives are achieved within a software application or system under development. SQA reviews span the entire lifecycle of a product, from requirements definition, software design, coding, code reviews, source code control, software configuration management, testing, release management, integration, provisioning, and deployment.

An SQA checklist indicates the standards, processes, and methods prescribed for the SQA review process. The SQA checklist shown on the next page implements the following sections and subtopics:

- **Project Management (PM):**
  - Procedural controls
  - Resources
  - Documented activities
  - Tracking and oversight
- **Methodology:**
  - Software methodology
  - Application or written controls
  - Technical reviews during development
  - Testing
  - Requirements information
  - Design information
  - Code listing
  - Performance and maintenance history
  - Purchased software products and service
  - Compatibility with bundled product
  - Virus free entry
  - Hardware methodology
  - Application of written controls
  - Technical reviews
  - Testing
  - Purchased hardware products and service
  - Compatibility with bundled product
- **Testing:**
  - Procedural controls
  - Test document and structure
  - Testing in the user environment
  - Software
  - Product maintenance
  - User manual
- **Configuration Management:**
  - Planned and user activity
  - Change and version management
  - Status of Change Management items
- **Documentation and Records Management:**
  - Documents that direct work activity
  - Technical and user documentation
  - Supporting record
  - Electronic documents and records.

System Quality Assurance Checklist
Project Name
Version

1 Project Management (PM)

System Topics	Yes/No	Comments
Procedural Controls		
• Project planning organizational policy is available.		
• Procedures available for establishing and reviewing project plans, commitments, complexity, costs, efforts, resources, facilities, personnel assignments.		
• Procedures available for work breakdown.		
Resources		
• PM assigned.		
• PM can negotiate work commitments.		
• PM trained in project planning and estimating.		
Documented Activities		
• Project activities documented, e.g., statement of work, project plans, resource estimates, risk assessment, faculties, work breakdown structures, metrics, project status, control, management, and contract reviews, etc.		
Tracking and Oversight		
• Policies exist to track project and oversight.		
• Project uses procedures to track actual time, funding, costs, and QA results, work size and complexity, scope, corrective action, changes, commitments, agreements, plans, schedules, risk, and resources.		

WEBSITE TESTING SUMMARY TEMPLATE

In modern systems development practices, web-based applications have similar development and testing complexities, size and scope of work observed in the traditional client-server and mainframe applications. Given their size and complexity, modern websites require the same kind of test planning that other software applications and systems require. The primary differences are in the standards, protocols, and technologies employed in the development of the web-based applications.

The Website Testing Summary Template provides summary information with QA and usability checklists to guide website quality assurance testing. Each checklist table provides questions or statements for which the tester responds with a Yes/No answer and respective comments where applicable.

The Website Testing Summary Template shown on the following pages incorporates more than 100 checklist items, spanning the sections and subtopics included in the list below:

- **General Website QA Review**
  - Site Design
  - Page Design
  - Navigation
  - Search
  - Text and Typography
  - Links
  - Graphics
- **Web Testing**
  - Functionality Testing
  - Usability Testing
  - Interface Testing
  - Compatibility Testing
  - Performance Testing
  - Security Testing
- **Usability Checklist**
  - Clarity of Communications
  - Accessibility
  - Consistency
  - Navigation
  - Design & Maintenance
  - Visual Presentation
  - Web Techniques to Avoid

YOUR COMPANY LOGO

Website Testing Summary Template
Project Name
Version

1 Overview

This document provides summary information and checklists for web quality assurance testing. Each checklist table provides questions or statements for which the tester responds with a Yes/No answer and respective comments where applicable. Completion of the checklists will help ensure the applications, functions, or features meet adequate quality assurance before being moved to production for end-user utilization.

Note: This document does not take the place of respective information documents that require full and complete tests, which may include test plans, test cases, scripts, etc.

2 General Website QA Review

System Topics	Yes/No	Comments
Site Design		
• Logo has a tagline briefly explaining what the company or application does.		
• Content is useful and relevant to the audience.		
• Content is clear and concise.		
• Information is grouped into logical units.		
• A privacy policy is provided.		
• A user feedback or problem reported capability is available.		
• The site has been tested on actual users.		
Page Design		
• Each page has a title.		
• Logo is on top left corner of the page.		
• Horizontal scrolling is avoided or minimized.		
• Layout is consistent throughout the site.		
• Ample white space is used.		
• Long pieces of content, such as articles, begin with a short summary at the top of the page.		
• Hyperlinks within content are used to point to other relevant content.		
• A consistent footer appears on every page of the site.		

SYSTEM TEST PLAN

The system test plan document is similar to the test plan template. The primary difference is one of scale. The test plan is generic and useful on most small to medium sized product development efforts. The systems design plan is oriented to testing more complex systems that involve additional integrated components.

The system test plan documents all system requirements identified in the functional and non-functional requirements specifications and design documentation. The objective of the system test plan is to identify and execute tests that fully exercise the system across all identified requirements.

This system test plan template helps document the strategies necessary to verify that a product conforms to its specifications and other requirements. The system test plan supports the following goals and objectives:

- **Identifies functions and features to be tested**
- **Documents the testing approach**
- **Identifies resources and schedules**
- **Identifies quality metrics defining "what good looks like"**

The System Test Plan contains the following sections and subtopics:

- **Purpose**
  - Reference Documents
- **System Test Description**
  - Test Goals and Objectives
  - Test Entrance and Exit Criteria
    - Entrance Criteria
    - Exit Criteria
- **System Test Approach**
  - Scope of System Testing
    - Test Categories
  - Risk Assessment
- **Test Environment**
  - Hardware
  - Software
  - Tools
- **Test Plan Schedule**
- **Testing Matrix**
  - Assumptions, Pre-Conditions, Risks
  - Test Instructions
  - Test Completion Summary
  - Associated Defects

System Test Plan
Project Name
Version

3 System Test Approach

This section provides information about the system testing environment that will be used, which should replicate the production environment. System testing includes test scripts based on the requirements, design, and specifications documentation.

3.1 Scope of System Testing

Provide information about what is and what is not part of system testing.

3.1.1 Test Categories

List the tests that will be performed as part of system testing.
Note: Add categories not listed and remove categories that will not be tested.

Category	Description
Functionality	*Functions as described in documentation.*
Security and Access Control	*Provides the proper application and user level security.* • *Application-level security, including access to the Data or Business Functions* • *System-level security, including logging into or remote access to the system.*
Data, Database, and Data Integration	*Ensure data accessed, used, and applied is valid and correct. Test databases and the database processes as a subsystem.*
Boundaries	*Fields perform in accordance with the constraints placed on the fields.*
Audit Trail	*Can track user and system activity (adds, changes, and deletes).*
Error Conditions	*Provides specific confirmation and error messages.*
Performance	*Provides or meets required performance guidelines as described in documentation.*
External Interfaces	*Functions timely and correctly with other external systems.*
User Interface	*User interfaces functions as described in documentation.*
Reporting	*Prints or displays report data as described in documentation.*

USER ACCEPTANCE TEST PLAN (UAT) TEMPLATE

In the traditional software development model, users only get once chance to review the software and agree whether it meets their requirements. The User Acceptance Testing (UAT) Plan is the planned activity that gives them that opportunity to assess the software. UA tests implement test scenarios that allow users to test the features and functionality of software application in support of real-world conditions and processes. The UAT approach provides a much more realistic assessment of the software in support of the users' day-to-day activities.

The UAT template provides an overview of the system, applications, functions and features that are to be tested in the User Acceptance Test Plan (UAT) process. Most importantly, the UAT plan should provide multiple test scenarios, each with associated test cases, that provide precise guidance on what tests need to be performed, and how the test will be performed.

Utilization of the UAT accomplishes many goals, but most importantly, it measures the quality of the application from the perspective of the customers and end users in context with supporting their business and operational needs.

The UAT template includes the following sections and subtopics:

- **Purpose**
  - Reference Documents
- **User Acceptance Test Description**
  - Test Goals and Objectives
  - Test Entrance and Exit Criteria
    - Entrance Criteria
    - Exit Criteria
    - Test Deliverables
- **UAT Test Approach**
  - Scope of UAT Testing
  - Test Categories
  - Risks, Dependencies, Assumptions, and Constraints
- **Functional Testing**
  - Functionality Included
  - Functionality Excluded
- **Test Environment**
  - Hardware
  - Software
  - Tools
- **Test Plan Schedule**
  - Roles and Responsibilities
- **Testing Matrix**
  - Assumptions, Pre-conditions, Risks
  - Test Instructions
  - Test Completion Summary
  - Associated Defects

User Acceptance Test Plan (UAT)
Project Name
Version

Note: Text displayed in blue italics is included to provide guidance to the author and should be deleted before publishing the document. In any table, select and delete any blue line text; then click Home→Styles and select "Table Text" to restore the cells to the default value.

1 Purpose

The purpose of the User Acceptance Test (UAT) Plan is to provide management an overview of the system, applications, functions, and features that are to be tested in the UAT process. Detailed information is outlined in the requirements, specifications, and design documentation. The plan and tests will provide information and guidance to management, staff, and the user community that the application works as expected and ensures a high level of confidence to implement.

It supports the following goals and objectives, which will help to verify the following:

- Functions, features, and items to be tested.
- The testing approach.
- Resources to be used and estimated testing time.
- The system can be used to perform the required business functions and processes under conditions that closely mirror the production environment.
- The system performs correctly as planned without error.
- System performance is acceptable.
- All requirements have been met through traceability from the documented requirements to the UAT scripts.

1.1 Background

Testers have been using test cases since the inception of computer programming over 50 years ago. The difficult part of creating test cases is determining which processing events should be made into test cases. Experience has shown that it is uneconomical to test all cases in an application system. The Return On Investment (ROI) is not worth the effort. Experience further shows that most testing exercises less than one-half of the total of all computer instructions. Therefore, it is mandatory that selecting the most important processing events is the key ingredient in building test cases.

1.1.1 Building Test Cases

The recommended process for the creation and use of test cases should follow the guidelines below:

Identify Test Resources	Testing use cases can be as extensive or limited a process normally dictated by time and budget constraints. Unfortunately, many testers approach the creation of test cases under duress and attempt to "catch" the most critical processing steps. Where their allotted time has expired, testing somehow is complete. One must bear in mind the actual time that has been allocated to conduct testing and then a process developed that optimizes that time.
Identify Conditions	Development of a testing matrix is recommended as the basis for identifying

TESTING BUG LIST

The Testing Bug List is another version of the Testing Error log, with a focus on strictly identifying and providing details on bugs.

The testing bug list provides a status of all identified bugs as they work their way through the bug fix process and should contain, at a minimum, the following columns and records:

- Bug reporting date
- Test Case ID (if applicable)
- Bug ID
- Bug name
- Bug description
- Severity level (High. Medium, low)
- Bug Status (New, open, closed, pending, completed, on hold)
- Tester name
- Date tested
- Test procedure method of testing.

YOUR COMPANY LOGO

Testing Bug List
Project Name
Version

Note: Text displayed in blue italics is included to provide guidance to the author and should be deleted before publishing the document. In any table, select and delete any blue line text; then click Home→Styles and select "Table Text" to restore the cells to the default value.

Bug Reporting Date	Test Case ID	Bug ID	Bug Name	Bug Description	Severity Level	Bug Status	Tester Name	Date Tested	Test Procedure
99/99/99	*Tst-044*	*B-09*	*Short Name*	*Provide a description of the Issue.*	*High Medium Low*	*New Open Closed Pending Completed On-Hold*	*R. Tester*	*99/99/99*	*Method of Testing*

Confidential – ©2015 Documentation Consultants (www.SDLCforms.com) Document: 5400 Page 1 of 1

www.SDLCforms.com

TESTING BUG REPORT

Many development organizations and project teams prefer a more detailed and formbased record of identified bugs and defects. The testing bug report shown on the next page provides an example of the report format.

The report includes the name of each bug/ defect – usually related to the name of the requirement not correctly implemented, area description, bug description, severity, status, priority, tester name, date tested, environment, test manager and tester names, method of testing, and who the bug was assigned to.

The Testing Bug Report shown on the next page contains the following data fields: • Reporting Date

- Test case ID
- Bug ID
- Bug Name
- Bug Description
- Area Path (e.g., Home Page/Login/Create New User)
- List of team members
- Build Number
- Severity Level (H, M, L)
- Priority (H, M, L)
- Tester Name/ Phone Number
- Test Manager/ Phone Number
- Date Tested
- Assigned To
- Bug Status
- Test procedure
- Environment

YOUR COMPANY LOGO

Testing Bug Report
Project Name
Version

Note: Text displayed in blue italics is included to provide guidance to the author and should be deleted before publishing the document. In any table, select and delete any blue line text; then click Home→Styles and select "Table Text" to restore the cells to the default value.

Reporting Date	*99/99/99*								
Test Case ID	*Tst-044*								
Bug ID	*B-09*								
Bug Name	*Application crashed while clicking on the Save button when creating a new user.*								
Bug Description	*The application crashed while clicking on the SAVE button while creating a new user; hence, I was unable to create the new user. Error Message 2013-09 displayed on the screen.*								
Area Path	*Home Page→Login→Create New User*								
Build Number	*Version 56.09.01*								
Severity Level	☐	**High**	☐	**Medium**	☐	**Low**			
Priority	☐	**High**	☐	**Medium**	☐	**Low**			
Tester Name	*B. Tester*	**Phone Number**	*999-999-9999*						
Test Manager	*R. Tester*	**Phone Number**	*999-999-9999*						
Date Tested	*99/99/99*								
Assigned To	*Bill Developer*								
Bug Status	☐	New	☐	Open	☐	Closed	☐	Pending	☐ Completed
Test Procedure	*Method of testing*								
Environment	*Windows 2010/SQL Server 2010*								

REGRESSION TESTING PLAN TEMPLATE

By definition, regressions are situations where the software capabilities have moved backwards (e.g., regressed) from previous implementation capabilities. This can happen when developers introduce bugs when enhancing, patching, or making modification to existing code. In addition, software and system configuration and environmental changes can cause regressions.

Regression testing evaluates software code, modules and applications to check for such regression, and simply requires re-running the functional and non-functional tests to ensure the acceptance criterion are still achieved for each affected requirement. The Functional testing sections identify the major activities, techniques, assumptions, constraints, and tools to be used to test the software applications. A Test Plan Schedule identifies each task description, number of days allocated for testing, and the start and end dates.

The regression testing plan template documents general information about systems or applications that require regression testing, including why the testing is required, functional business areas affected, and testing timeline.

The Regression Testing Plan includes a definition of regression testing, types of regression testing, the regression test approach, scope of regression testing, test categories, and risks, dependencies, assumptions and constraints.

The Regression Testing Plan contains the following sections and subtopics:

- **Introduction**
  - Reference Documents (e.g., Requirements & Technical Specifications)
- **What is Regression Testing**
  - Purpose
  - Types of Regression Testing
    - Functional and Unit testing
  - Regression Test Approach
    - Guidelines for patch/ Upgrade Releases
  - Scope of Regression Testing
  - Test Categories
  - Risk, Dependencies, Assumptions, and Constraints
- **Functional Testing**
- **Test Environment**
- **Regression Test Results**
- **Test Plan Schedule**
- **Testing Matrix**
  - Test Instructions
  - Test Completion Summary
  - Associated Defects

YOUR COMPANY LOGO

Regression Testing Plan
Project Name
Version

2 What is Regression Testing

2.1 Purpose

The purpose of Regression Testing is to seek to uncover new errors, or regressions, in existing functionality after changes have been made to a system, such as functional enhancements, patches, or configuration changes.

Definition: Regression testing is selective retesting of the system; executed with an objective to ensure the enhancement/bug fixes work and those enhancement/bug fixes have not caused any un-intended effects in the system. The intent of regression testing is to ensure that a change did not introduce new faults.

Common methods of regression testing include rerunning previously run tests and checking whether program behavior has changed and whether previously fixed faults have re-emerged. Regression testing can be used to test a system efficiently by systematically selecting the appropriate minimum set of tests needed to adequately cover a particular change.

2.2 Types of Regression Testing

There are 3 main types of regression testing:

1 Incremental Testing.
   - Test only the code that was added or modified against anticipated results.
   - Repeatedly test in a scheduled set of increments, e.g., after the nightly build.

2 Incremental Regression Testing.
   - Selective retesting of portions of the software that has changed, or is impacted by a change, against a known baseline of results to verify that there are no unintended side effects.

3 Full Regression Testing.
   - Complete retesting of the entire software application against a known baseline of results to verify that there are no unintended side effects anywhere in the application.

This page is intentionally left blank

Project Administration Phase
Overview

The project administration phase encompasses project management related activities to execute, monitor and control a project in accordance with the project plan. Ideally, each project management requirements have a method and a tool that facilitates the work during the project's duration. So, for example, if your project plan has a report plan that specifies monthly executive management communications of project performance against the approved constraints, a report plan template would specify the frequency, audience, the activities required to collect the necessary data, and the format in which the information is reported.

It is important to keep an accurate record of changes that have occurred to the baseline over the life of the project, whether they were formally approved or not. The records of such changes feed the lessons learned identified over the course of the project and help improve the accuracy of planning and estimating on subsequent projects.

Tools that should be utilized and continually updated include:

- Change Management Log - to record changes to project baseline information
- Action Item Status Log - containing action items identified and assigned on an ad hoc manner
- Risk Management Log – to identify risks, record risk mitigation strategies, project impact, probability of occurrence, timeline, the response and action completed to resolve any risks before they become a threat to completion of the project within budget and schedule
- Issues Identification and Resolution – to track problems when problems occur, and document contingency plans executed to limit the impact
- Project Milestone Status Form – to identify all project milestones and update changes to the schedules of deliverables and related milestone
- Project Status Report – to provide management updates in context with the customer or executive management approved constraints of the project. e.g., scope, budget, time, resources, and quality
- Meeting Summary – a tool to record substantive information discussed during a meeting, decisions made, and risk or issues identified or resolves, and any action items that were identified or resolved during the meeting

Forms & Templates

CHANGE MANAGEMENT LOG

Under the traditional software development model, change management is a critical concern. The plan driven and linear sequential processes created a situation in which developers and project managers were reluctant to change given the potential negative impact on the approved constraints of the project. (e.g., scope, time, budgets, resources, and quality)

The agile-based software development model effectively gets rid of the need for a change management process and log. Agile practices incorporate frequent and short development iterations with continuous customer and user feedback between iterations. The feedback from customers and users helps inform the development team on changes in requirements and new priorities. Therefore, change requests under the agile model are handled within the product backlog.

However, under the traditional software development model, change requests have impact on the projects plan, and the approved constraints. Therefore, the change request must be evaluated to determine the impacts and benefits, and the change request must be approved by management before allowing the baseline plans to be changed. After reviewing a change request, the executive sponsors or their designates may choose to authorize the new scope of work and other necessary changes to budgets, schedules and resources, or they may defer the work to future product release. In either case, the change management log keeps a record of these decisions.

Specifically, the change management log uniquely identifies and provides a record of all requested changes over the course of the project. The change management log uses a worksheet format and implements columnar formats with individual records for each change request. The typical fields within the change management log include:

- Change Request Number
- Change Type
- Change Description
- Priority (critical, high, medium, low)
- Request Owner
- Date Submitted
- Level of escalation
- Action or Resolution
- Date of Change Approval.

YOUR COMPANY LOGO

Change Management Log
Project Name
Version

Note: Text displayed in blue italics is included to provide guidance to the author and should be deleted before publishing the document. In any table, select and delete any blue line text; then click Home→Styles and select "Table Text" to restore the cells to the default value.

Change No.	Change Type	Description	Priority	Owner	Date Sub-mitted	Escala-tion Required	Date App-roved	Action / Resolution
CM-001	*Req'ts Scope Design Schedule Funding, etc.*	*Revised performance requirement that access time for web application be decreased from 3.4 seconds to 2.9 seconds.*	*Critical High Medium low*	*B. Parks*	*01-01-16*	*Yes*	*N/A*	*The Change Control Board denied the change on 01-12-16 as the impact to revise numerous modules at this late stage of development would have severe cost and schedule increases.*

 Document: 3010 Page 1 of 1

www.SDLCforms.com

ACTION ITEM (STATUS) LOG

Under the traditional software development model, project teams may experience situations where new tasks are identified in an ad hoc manner. If the identified tasks are not directly related to the activities defined within the project plan, or the impact against the project's constraints is minimal, it often easier to maintain a list of the actions in a separate log, as opposed to updating the schedule plan. Adding tasks to the schedule plan usually requires the establishment of a new baseline, forcing burdensome rescheduling, reviews, and approvals.

Working tasks against an action item log is a much simpler process. The ad hoc tasks can be identified at any time during the project and brought directly to the to the project managers attention or detailed during a project status meeting. The project manager documents the task in the action item log, but must discuss the task with the team members, and affected business process owners, managers, and executive sponsors to obtain consensus on the need and the impact.

The Action Item Log is typically managed in a worksheet format and application, and contains the flowing information:

- Task Name
- Task ID
- Task Description
- Task Impact
- Date assigned
- Task Owner
- Priority (high, medium, low)
- Task Completion Due Date
- Current Status
- Date Completed/ Resolved.

Action Item Status
Project Name
Version

Note: Text displayed in blue italics is included to provide guidance to the author and should be deleted before publishing the document. In any table, select and delete any blue line text; then click Home→Styles and select "Table Text" to restore the cells to the default value.

Item #	Description and Impact on Project	Date Assigned	Owner	Priority (H/M/L)	Due Date	Status	Date Resolved

RISK MANAGEMENT LOG

Project risks are identified throughout the life cycle of a project. Risks represent a potential danger to the project, usually affecting one or more of the projects constraints. In other words, the risk may impact the deliverables, scope of work, budgets, schedules, resources or quality of work – or any subset or all of the project constraints.

Since a risk, by definition, does not represent an immediate threat to the project, it's important that a record of identified is kept, and periodically reviewed, so that project team members and other stakeholders are aware of the potential and are constantly looking to see if the risk is becoming a threat. The risk management log provides a method and tool to maintain project's the risk record.

The Risk Management Log is usually implemented in a worksheet format and contains columns with records to record risk information. The risk log identifies the risk, who found it, who's responsible for developing the mitigation strategies, identifies potential the mitigation strategies, and records estimated measures for potential impacts and probability of occurrence.

The Risk Management Log shown on the following page contains the following sections and data fields:

- **Initial Risk Information**
  - Risk Description / Risk Event Statement
  - Responsibility for the risk

Date reported and date updated

- **Risk Assessment Data**
  - Impact (High / Medium / Low)
  - Specific impact of risk
  - Probability of occurrence (High / Medium / Low)
  - Timeline (Near term / Medium / Far Term)
  - Status of response
  - Plan developed but not enacted
  - Plan enacted but effect not determined
  - Plan enacted and effective
- **Risk Response / Action**
  - Completed action
  - Planned future actions
  - Risk status.

YOUR COMPANY LOGO

Risk Management Log
Project Name
Version

1. INITIAL RISK INFORMATION

Enter Unique ID for Risk Number	*A risk event statement may be what might happen in the future and has possible impact on the project. "Weather" is not a valid example of an event, but "Bad Weather" is.*	*Enter the name and/or title of the team member responsible for the risk.*	*Enter the date the risk was identified.*	*Enter the date the risk was updated.*
Risk Number	**Risk Description / Risk Event Statement**	**Responsibility**	**Date Reported**	**Last Update**
R-01	*Price changes are volatile and can change on a weekly basis, which will adversely affect profit calculations.*	*Price Group Manager*	*1/23/2016*	*2/12/2016*
R-02	*Key vendor on Supplementary Invoicing Package is experiencing cash flow problems.*	*IT Vendor Management*	*2/16/2016*	--

www.SDLCforms.com

ISSUE IDENTIFICATION AND RESOLUTION FORM.

Issues are problems that are already impacting the project. The issue may have been a previously identified risk. In such cases, the ideal situation is to have previously identified risk mitigation strategies to reduce the possibility of occurring or the impact if it does. But, if the risk mitigation strategies do not work, then the fall back is to have previously developed a contingency plan that lessons the impact if the risk turns into an issue. Unfortunately, organizations do not always do the risk mitigation and contingency planning they should. Also, sometimes, bad things happen regardless.

When an issue arises that was unanticipated, the project team must immediately analyze the problem to see what their resolution options are, and then execute the best alternative to fix the problem. That is the purpose of the Issue Identification and Resolution form.

This document is used to individually identify each issue that may impact a project, and identify who created and resolved the issue, the type of issue, potential alternatives and recommendations, provide an estimate of the resources, man hours and costs, and management actions that were taken to resolve the issue.

The Issue Identification and Resolution contains the following sections:

- Purpose
- Issue identification
- Issue type
- Complete description of issue
- Potential impact if issue not resolved
- Assignment of issue
- Resolution alternatives
- Resolution recommendations
- Resources required
- Management action to resolve issue

YOUR COMPANY LOGO

Issue ID & Management Resolution
Project Name
Version

Note: Text displayed in blue italics is included to provide guidance to the author and should be deleted before publishing the document. In any table, select and delete any blue line text; then click Home→Styles and select "Table Text" to restore the cells to the default value.

1 Purpose

This document is used to individually identify each issue that may impact a project, and identify who created and resolved the issue, the type of issue, potential alternatives and recommendations, provide an estimate of the resources, man hours and costs, and management actions that were taken to resolve the issue.

After identifying the issue, record critical parameters in the Issue Management Log for action.

2 Issue Identification

This section is where you identify the issue name and who originated the issue.

Issue Name:			
Originating Dept.:			
Responsible Dept.:			
Issue Created By:		**Date Issue Created:**	
Issue Number:		**Date Resolution Required:**	
Project Manager:			

3 Issue Type

Fill in the following table with the appropriate information.

Technical				
☐ Development	☐ Infrastructure	☐ Database (DB)	☐ IT Governance	☐ Prod. Support
☐ Other				
Non-Technical				
☐ Business Unit	☐ Project Planning	☐ Funding Resources	☐ Personnel Resources	
☐ Other				

 Document: 6150 Page 4 of 6

ISSUES MANAGEMENT LOG

The Issues Management Log implements a method as a worksheet-based tool to maintain a record of identified issues. Issues are problems that impact the project in some way, typically affecting the constraints of the project.

The Issue Management Log consists of 3 sections to document each issue:

1. Initial Issue Information

2. Issue Assessment Data

3. Issue Response / Action.

The Initial Issue Information section contains a log, similar to the risk management log, to record the issue number, enter the Risk ID number (if applicable), provide a brief description of the issue, document who is responsible for addressing the issue, record the priority and impact severity metrics, and the date it was opened.

The Issue Assessment section identifies the impact, the project or product area that's impacted by the issue (development, database, etc.) and the target resolution date.

The Issue Response / Action section identifies the resources required to resolve the issue, document identified alternatives and recommendations, and the action taken by management.

Issues Management Log
Project Name
Version

YOUR COMPANY LOGO

1. INITIAL ISSUE INFORMATION

Enter Unique ID for Issue Number	*Enter Risk ID (if from Risk Mgt. Log)*	*Provide a complete description of the issue.*	*Enter the name and/or title of the team member responsible for the issue.*	*Enter the priority as: High Medium Low*	*Enter the date the issue was identified.*
Issue Number	**Risk Number**	**Issue Description**	**Responsibility**	**Priority (H/M/L)**	**Date Opened**
I-01	*R-01*	*Price changes, which are volatile and can change on a weekly basis, are embedded in the programming code and will adversely affect profit calculations if not continuously updated.*	*Price Group Manager*	*Medium*	*2/12/2016*
I-02	*R-02*	*Key vendor on Supplementary Invoicing Package is experiencing cash flow problems.*	*IT Vendor Management*	*High*	*2/19/2016*

www.SDLCforms.com

PROJECT MILESTONE STATUS FORM

Milestones are project markers of important events or deliveries in the project schedule. Milestones have zero time associated with the events as they represent scheduled markers in time, and not a recording of work accomplishments. Think of milestones as planned achievements or accomplishments. When milestones are completed on time the project is advancing properly in accordance with the original schedule.

Scheduled milestones may have milestone review meetings associated with them. The milestone meetings give the customers and executive sponsors team an opportunity get into the details on the status of the project.

It's not unusual to schedule milestone reviews at the completion of the phase of work or the beginning of a new phase of work. Such milestone reviews typically have detailed checklists that the project and development team complete in preparation for the meeting. The checklists provide lists of critical tasks and deliverables associated with the phase of work under review. The purpose of the checklists is to ensure the identified work is completed or identified as not applicable. In other words, the checklists provide a quality assurance check to ensure all necessary work is completed before asking approval to close the completed work or start the new phase of work.

This Project Milestone Status Form implements a method and provides a worksheet-based tool to identify project milestones and record their status as the progress progresses. The worksheet documents the name of each milestone, the milestone ID numbers, purpose, responsibility assignments, due dates and completion dates, and the then current status of each milestone.

The Milestone Status Form shown on the next page contains the following sections and data fields:

- Project general description
- Project Manager
- List of team members
- Milestone ID Numbers
- Definition of milestone, goal, function or task
- Responsible Person
- Due Dates
- Completion Dates
- Milestone Status (in-process, completed or delinquent).

Project Milestone Status Form
Project Name
Version

Note: Text displayed in blue italics is included to provide guidance to the author and should be deleted before publishing the document. In any table, select and delete any blue line text; then click Home→Styles and select "Table Text" to restore the cells to the default value.

Project General Description		Project Manager		
		Team Members		

ID #	Milestone, Goal, Function or Task	Responsible Person	Due Date	Compl-etion Date	Milestone Status		
					In-Process	Com-pleted	Delin-quent

Confidential – ©2015 Documentation Consultants (www.SDLCforms.com) Document: 6250 Page 1 of 1

www.SDLCforms.com

PROJECT STATUS REPORT

Project status reporting is at the heart of project management in the traditional software development model. Through the years, the reporting requirements have not changed a great deal. The primary components of a project status report include project summary information, schedule information, status on deliverables, important measurements indicating performance against the constraints of the project, list of accomplishments and upcoming tasks from the project schedule, identified risks and issues, change requests, and action items. Other items can be added as necessary.

Project status reports are developed and submitted fairly frequently, usually once every week or every two weeks. Project Status Meetings are typically held with the project's executive sponsor, line managers, program managers or IT managers to present and discuss the details contained within the reports.

The Project Status Report shown on the following page contains the following sections:

- **Distribution**
- **General Overview of Project Status**
- **Administration**
- **Project Activity**
- **Problems or Delays Encountered**
- **Issues to be Resolved**
- **Planned Activities**

In addition, the report provides additional details on the deliverable items:

- **Deliverables**
  - Deliverable Descriptions
  - Planned Date
  - Actual Date
  - Status

Other status items are added to this report on a discretionary basis, such as.

- Project Schedule Tasks – Accomplished
- Project Schedule Tasks – Upcoming
- Project Schedule Tasks – Past Due
- Graphical Headlight (Red, Yellow, Green) show status of project constraints
- Change Requests (Open, Approved, Rejected)
- Issues Log
- Risk Log
- Action Items

YOUR COMPANY LOGO

Project Status Report
Project Name
Version

Note: Text displayed in blue italics is included to provide guidance to the author and should be deleted before publishing the document. In any table, select and delete any blue line text; then click Home→Styles and select "Table Text" to restore the cells to the default value.

To:	
cc:	
From:	
Date:	
Subject:	<Project Name> status from <dd-mm-yyyy – dd-mm-yyyy>

General
Provide a project status overview for the respective time period.
Administration
Provide information about any staff, communication, and administration changes that can impact the schedule or project.
Project Activity
Provide relevant activity information about the project, e.g., analysis, requirements, design, architecture, development, testing, and implementation.
Problems or Delays Encountered
Provide any information about project activity problems or delays.
Issues to be Resolved
Provide any information about issues to be resolved, including suggested resolutions or solutions under review.
Planned Activities Next Period
Provide any information about planned activities for the next status report time frame.

MEETING SUMMARY TEMPLATE

Meetings occur in both agile and traditional software development projects. It's often important to keep a record of items discussed, decisions made, and additional action items identified. The Meeting Summary template provides a tool to document the meeting date and time, participants, meeting minutes, conclusions, and action items status.

The Meeting Summary template contains the following sections and subtopics:

- Meeting Subject
    - Meeting Originator
    - Meeting Date / Time
    - Attendees
- Meeting Overview
- Discussion about meeting topics
- Any conclusions
- Identification of potential issues or risks
- Action item status: the status, description, person responsible, due date and any comments.

YOUR COMPANY LOGO

Meeting Summary
Project Name

Meeting Subject				
Meeting Originator				
Meeting Date / Time				
Attendees				

Meeting Overview

Discussions

Conclusions

Issues / Risks

Attachments	

Status	Action items	Person Responsible	Due by	Comments

www.SDLCforms.com

Deployment Phase

Overview

This phase of work includes activities spanning deployment of the solution into pre-production testing and production environments, plus turnover to the product support, maintenance, and sustainment groups. Although the ultimate goal is to get the software or system into the hands of customers and users, deployment phases processes include pre-deployment, deployment and post-deployment activities. Deploying a product, particularly an enterprise application, is an involved set of processes that begins well before a product is ready to be or installed at a customer site or accessed via a network and continues on for some time after delivery.

- Typical pre-deployment activities include:
- Communications – Customers and end users need to know when to expect an upcoming release or new product and how the release will affect their organizations. If the application affects business processes, forthcoming changes to the process needs to be communicated.
- Training Artifacts - Users and operations staff may require training on how to use, support and maintain the system, and training aids and courses must be developed in advance of the deployment.
- Configuration Management - The configuration managers need to ensure all code, components, documentation and other artifacts associated with the release are properly identified with the proper version and archived for future reference.
- Reference Guides – Develop and distribute guides, in either document or electronic form, for installation, system administration, system operations, product support and end-user references.
- Procurement – Necessary computing equipment, network systems, data stores, security components, and software resources must be identified and procured in advance of the solution's delivery.

Rollback Plan – In case the system fails when put into production, the rollback plan defines the process for recovering and restoring data and cutting back over to the original software or systems.

- When the product is ready for release to the customer or into the production environments, the release manager ensures the pre-deployment activities are completed and then oversees the execution of the release.
- Provisioning – Shipment and delivery or network access to computing equipment, networks, data stores, security components, and software resources.
- Installation & Configuration – Ensuring all hardware and software is properly installed, configured, integrated, and tested in the production environment.

Training – deliver training artifacts, WIKIs, and hand-on training sessions, as required. Also, ensure all operations and user guides are deployed with the system.

- Once the system is deployed into the production environment, the development team has a period of time in which they remain involved. Some of the common activities they provide include:

- Monitor –To ensure the product works as designed, and to address any issues that arise.

- Bug Fixes – if the customers observe bugs that effect the functionality of the system, the development team may be called upon to fix them before the final turn-over of the system.

- Change Requests – as customers work with the system, inevitably they will find new capabilities that they would like the system to implement. Change requests are a formal way to request those changes, and even under agile practices, may be required if funding is not available to continue development on the system.

- Product Acceptance – After the Warranty period, negotiated in advance of deployment, the customer reviews the status of the deliverables and conformance with requirements. Assuming everything was delivered properly, the client signs the turnover approval forms, verifying their acceptance.

Note that *Release Management* is another common term used in deployment. As an example, the ITIL-based release and deployment management process encompasses planning, scheduling and controlling the movement of product releases (including hardware, software, documentation, business process changes, and other associated release/ configuration items) to the test, pre-production and production environments.

- This chapter includes templates for developing necessary Guides used after the system is deployed into the organization's or customer's production environments. The Guide Templates include:

- Process Guide Template

- Installation Planning Guide Template

- Software user Guide Template

- System Administration Guide Template

- Operations Guide Template

Regardless of the terms used to describe this phase of work, the deployment activities are necessary to support both agile and traditional software development practices. Larger and more complex systems require more planning and preparation in advance of their releases. The Production Implementation Plan (PIP) supports deployment planning. The PIP is especially useful when a product must be deployed across a number of disparate environments maintained by different organizations.

Finally, the production turnover approval process and final signoff form is required to ensure the receiving organizations have agreed the solution is operating as intended, that their sustainment organizations are trained and ready to assume responsibility for maintaining the solutions, that their help desk organizations are properly trained and ready to support end-users, and to provide formal acknowledgment that the customers or end-user organizations have assumed full control over the solutions. This approval may not come until after a warranty period has expired to ensure the sustainment and supporting organizations have had adequate time for knowledge transfer, and that no critical bugs show up in the deployed solution.

Forms & Templates

PROCESS GUIDE TEMPLATE

Business systems often support, integrate and automate critical business processes. Those processes can be cross-functional and cross-organizational in scope. Before the supporting system can be designed and developed, the business analysts or development team members needs to document the details of the process.

If the process definition effort only looks and how the business process is currently executed, the organization is missing an opportunity to make it better. The automation of an existing process is rarely a good thing. If the process isn't efficient and streamlined and includes work that is not necessary or valued by the customer, then automation will simply make those efficiencies, bottlenecks, and non-value-added work more costly.

A process guide is used to identify, document and communicate how work is executed in a business process. In the traditional software development model, the process guide is developed through a series of information gathering sessions with the organizations process owners, process improvement teams (PIT), executives, managers and process participants.

In support of process documentation effort, the process guide template, shown on the next page, implements the so-called *"Five Ws and How"* questioning approach to information gathering and problem solving. The questions include:

Who – is responsible for each defined activity and task?

What – procedures must be followed for each defined activity and task?

When – is the activity or task performed?

Where - is the activity or task performed?

Why - is the activity or task performed?

How - is the activity or task performed?

The template also facilitates documentation of information about the system, application or process instructions, procedures, and workflows, which are shown in step-by-step text format as well as visual graphics to explain the action performed.

- **The Process Guide contains the following topics:**
  - Introduction
  - Purpose and Scope
  - Background
  - Audience
  - Referenced Documentation
  - Process Information
- **Major Procedures, Tasks, and Functions**
- **Other Process Information.**

Process Guide
Project Name
Version

2 Process Information

For each process, provide the following information (e.g., who, what, where, when, why, how) to explain the action performed:

- *Who is responsible for the activity or task, e.g., project approval and budget; who needs to sign off on a project; how many reviews are required and when?*
- *What procedures need to be followed to perform the task?*
- *Where is the activity or task performed?*
- *When does the phase, activity or task start and end?*
- *Why is the phase, activity or task performed?*
- *How is the activity or task performed?*

Notes:

- *Include graphics and process flows in this section or place them in the Appendix Section when more space is required.*
- *Copy and paste the tables in the Process Section for each process. Repeat it whenever necessary.*

Topic	Description
Location	*Business Unit, Department or Region to which the process applies.*
Process Name:	*Name / abbreviation for the process.*
Parent Activity / Process	*Parent business activity or process. If none, enter N/A.*
Description	*Describe the main purpose of the process.*
Primary Product	*Identify the primary product(s) with a brief explanation for each. Who will use each primary product?*
Trigger	*Identify the event(s) that trigger the process. Triggers include events such as dates, tasks, or external processes.*
Sub-processes	*When the main process is sub-divided, Indicate each sub-process with a brief description.*
Standard Path Events / Activities	*Identify the activities and events that occur for the main process.* *Identify an activity and event that occurs in a specific sub-process.* *Highlight where and how an alternative path breaks off from the main path.*

INSTALLATION PLANNING GUIDE

Large, multi-tier, distributed, and highly integrated systems have complex installation and configuration requirements. Those installation and configuration procedures must be documented before the system is deployed to customers and internal operations support groups.

The primary purpose of the Installation Planning Guide is to capture and maintain information required to successfully install and configure an application. Specifically, the guide implements both a method and a tool to document information related to the installation of the system, application or data - including installation strategies, environmental concerns (e.g., hardware, software, networks, security, data stores, virtualization, failover), configuration and setup requirements, planning and risk factors, backup and restore procedures, and other important considerations..

An Installation Planning Guide is an essential document when the installation process is sufficiently complex to require a documented plan. (e.g., includes pretty much any software or system installation that can't be packaged in a network-based download or shipped on a CD.)

The Installation Planning Guide shown on the next page contains the following sections and subtopics:

- Introduction
  - Purpose
  - Objectives
  - Key Stakeholders and Points of Contact
  - Referenced Documentation
- Installation Information
  - System Overview
  - Installation Overview
  - Scope
  - Environment
  - Installation strategy
  - Installation risk factors
  - Security
  - Installation planning
  - Pre-installation plans and requirements
  - Installation schedule
- Installation Instructions
  - Major Phases
  - Tasks and Procedures
  - Backup Procedures
  - Rollback and Recovery Procedures
  - Change Control Procedures
- Installation Support
  - Hardware Inventory
  - Software inventory
  - Network Inventory
  - Facilities
  - Facilities
  - Materials
  - personnel

YOUR COMPANY LOGO

Installation Planning Guide
Project Name
Version

2.5 Installation Strategy

Describe the strategies to convert or install system hardware, software, and data.

- *Hardware Conversion or Installation Strategy: Includes the new hardware environment.*
- *Software Conversion or Installation Strategy. Includes the software strategy to be used.*
- *Data Conversion Strategy. Includes the strategy for data conversion, data quality assurance, and data conversion controls.*
- *Interfaces. Includes the new or affected interfaces. For a hardware platform conversion, e.g., mainframe to client/server, provide information about affected interfaces and their revisions.*
- *Data Quality Assurance and Control. Includes assurance for data quality before and after all data conversions, e.g., approach to data scrubbing and quality assessment of data before they are moved to the new or converted system.*

Strategy Topic	Description
Hardware	
Software	
Data Conversion Approach	
Interfaces	
Data Quality Assurance and Control	

2.6 Installation Risk Factors

Describe the major risk factors in the installation and strategies for their control or reduction, e.g.,

- *Technical performance of the installed system*
- *Installation and conversion schedule*
- *Costs*
- *Backup and recovery procedures.*

2.7 Security

Provide an overview of the system security features and security during the installation.

SOFTWARE USER GUIDE

An optimal goal is to develop software that is so intuitive the users can immediately start utilizing it and being productive without any use of reference guides or training. Unfortunately, that's not always possible. This is especially true for complex applications that support the organization's cross-functional and cross-departmental business operations. Yes, perhaps the users can figure out the use of the product's features and functions over time. But those activities are a form of waste that add costs and are not value added from the customer's perspective.

Software user guides provides reference information on how to use a software product or system from a contextual use perspective. In other words, the software user guide should provide information on how to use in support of the function of the product. The guide also describes the major components, user benefits, access information, navigation instructions, data, analytics, and reports.

To be an effective aid, those who develop the user guides should use common language with relatively short sentences. The software guides should contain both text-based information along with clarifying images and screenshots from the application.

Application navigation instructions should include descriptive text, screenshots of the graphical user interface (GUI) and procedural or workflow diagrams. The language used is matched to the intended audience, with technical terms kept to a minimum or explained thoroughly. In other words, use domain language, and not IT language – unless IT specialists are your intended audience.

The Software User Guide shown on the next page contains the following sections and subtopics:

- **Introduction**
  - Purpose and Scope
  - Background
  - Audience
  - Referenced Documentation
- **What Is the (Application, System or Process)**
  - Major Components, Functions, and Features
  - Benefits
  - Users
  - Authorization Profile
- **Access Information**
  - General Information A User Should Know
- **Navigation**
  - Navigating Menus
  - Navigate Main Summary
  - Action Item Description and Uses
- **Major Procedures, Tasks, and Functions**

YOUR COMPANY LOGO

Software User Guide
Project Name
Version

3 Access Information

3.1 General Information A User Should Know Before Getting Started

Provide information about how to login into the application, system, or process, e.g.,

- *Login with your User ID and password. The System will navigate to the home page, with a navigation menu on the left and a blank form on the right.*
- *Once you are registered into the application, system or process, your ID will be your username and your current password will be your password. There is no need to change this, changing it may cause difficulty later on.*
- *If you are not registered and require access, you must send a message to xyz@companyname.com (administrator) requesting use of the system.*
- *At the time of registration, you will be affiliated with the customer team(s) you support. Once these customer teams are selected, the only information you will be able to access will be for the customer teams you support. The customer team drop down menus located throughout will only reflect the customer teams you support.*
- *The System is case sensitive: Base Number XYZ-01-000123-A1 is not the same as xyz-01-0000123-a1 and will not enable you to access the document.*

 Document: 7100 Page 7 of 17

www.SDLCforms.com

SYSTEM ADMINISTRATION GUIDE

Systems administration guides, often shortened to "Sysadmin Guides," describe how a system is deployed, supported, and maintained by the IT operations organization. Procedures and information contained within the document span the following product support and maintenance activities:

- Installation and configuration
- Systems integration
- System startup and shutdown
- System backups
- System recovery
- Performance monitoring
- Use of transaction logs
- Use of utility commands
- Setting up user and group accounts
- Establishing user roles, privileges, access rights, and passwords
- Database support and maintenance activities
- Implementing security capabilities
- Provisioning and monitoring software licenses

The System Administration Guide is targeted at a technical audience, generally those who work within the IT department's operations group. As such, the verbiage used within the Sysadmin Guide is IT domain centric. Still, the system administration guides should not be overly complex and loaded with technical jargon that doesn't add value to the descriptions. The sysadmin guide should use text-based instructions with clarifying images and screenshots of the application.

The System Administration Guide shown on the following page contains the following sections and subtopics:

- Introduction
  - Purpose
  - Objectives
  - Referenced Documentation
- General System Information
  - Overview
  - Data Assets
  - Processing
  - Environment
    - Facilities
    - Hardware Inventory
    - Software Inventory
    - Network Inventory
- Administration and Maintenance
  - Server Administration
  - Account Administration
  - System & Software Administration and Maintenance
  - Database Administration and Maintenance
  - Backup

System Administration Guide
Project Name
Version

Note: Text displayed in blue italics is included to provide guidance to the author and should be deleted before publishing the document. In any table, select and delete any blue line text; then click Home→Styles and select "Table Text" to restore the cells to the default value.

1 Introduction

This document is a System Administration Guide, which provides procedures and information to administer and maintain the < Name of System >.

Note: Delete or indicate document sections that are not applicable.

1.1 Purpose

Provide information about the purpose of this system administration guide, including information about the system and its operation.

1.2 Objectives

Provide the business or technology objectives to be achieved with this system administration guide.

1.3 Referenced Documentation

Provide information about all documentation referenced in this document, e.g., number, title, version, and date.

Document Number	Version	Date	Document Name

OPERATIONS GUIDE

An operations guide is a complementary manual to the sysadmin guide. Where the sysadmin guide provides general technical instructions related to supporting and maintaining a deployed system, the operations guide gets much more into the details of how the systems is supported and maintained. Similar to the process guide template, the operations guide also implements the *Five W's and How* questioning approach to define how the system maintains an operational status on a day-to-day basis. (E.g. Who, What, When, Where, and How?)

The operations guide template shown on the following page includes the following sections and subtopics:

- **Company History, Vision & Organization**
- **Products & Services**
- **Policies**
- **Position Statements**
- **Systems (Action Plans)**

The Operations Guide contains the following topics:

- **Introduction**
  - Purpose and Objectives
  - Audience
- **General System Information**
  - Overview
  - Key Contacts
  - Environment and Assets
    - Facilities
    - Hardware Inventory
    - Software Inventory
    - Network Inventory
    - Interfaces
- **Operations, Administration, and Maintenance**
  - Processing Overview
  - Responsibility
  - Operations Calendar
  - Operating Procedures
- **Maintenance and Troubleshooting**
  - Change Management
  - Configuration Management
  - Server Administration
  - Account Administration
  - System & Software Administration and Maintenance
  - Database Administration and Maintenance

Operations Guide
Project Name
Version

Note: Text displayed in blue italics is included to provide guidance to the author and should be deleted before publishing the document. In any table, select and delete any blue line text; then click Home→Styles and select "Table Text" to restore the cells to the default value.

1 Introduction

This document is an Operations Guide, which provides procedures and information to administer and run the < Name of System >.

Note: Delete or indicate document sections that are not applicable.

1.1 Purpose and Objectives

Provide information about the purpose and objectives of this operations guide, including information about the system and its operation. This guide is an important source of information to support the following:

- *Daily, weekly, and monthly operations*
- *Special operations*
- *Unique tasks, e.g., hardware, software, license upgrades*
- *Troubleshooting*
- *Auditing*
- *Best practices and quality standards.*

1.2 Audience

Provide information about who the intended audience for this guide is.

1.3 Referenced Documentation

Provide information about all documentation referenced in this document, e.g., number, title, version, and date.

Document Number	Version	Date	Document Name

PRODUCTION IMPLEMENTATION PLAN (PIP)

The Production Implementation Plan provides detailed planning information for the production approval and implementation of the software product or system and describes the steps necessary to place the product into production. The PIP identifies the objectives, impacted devices, production delivery steps, technical support information, hardware and software components, testing and acceptance, rollback/contingency plan, and required training and documentation.

Its main purpose is to minimize if not eliminate any negative impacts to the production environments when a new or modified application is made available to install in the production environments, The PIP is often included as part of the transition plan governing how a new or modified application is moved from the development environment and into the production environment.

The Production Implementation Plan contains the following sections:

- **Production Implementation Plan**
  - PIP Description/Objective
    - Impacted Devices
    - Production Delivery
  - Technical Support Information
  - Potential Impacts
  - Software Components & Implementation Steps
  - Hardware Components
  - Hardware Implementation Steps
  - Testing and Acceptance
  - Rollback and Contingency
  - Required User Training and Documentation
  - Other Emergency Contacts.

YOUR COMPANY LOGO

Production Implementation Plan (PIP)
Project Name
Version

1 PRODUCTION IMPLEMENTATION PLAN DESCRIPTION

Note: Text displayed in blue italics is included to provide guidance to the author and should be deleted before publishing the document. In any table, select and delete any blue line text; then click Home→Styles and select "Table Text" to restore the cells to the default value.

This form provides planning information for production implementation of the <Name of Project>.

Topic	Description
Project Name	
Project Number	
Department or Business Unit Name	
Department / Business Unit Cost Center	
Sponsor Name and Telephone Number	
Project Manager Name and Telephone Number	

Objective: This production implementation plan (PIP) describes the final steps necessary to implement the < Name of Project >.
Impacted Devices: The following is a list of names and logical addresses for all impacted applications / systems / databases / network devices.
Production Delivery: The following table lists the general steps necessary to deploy the product to production. Any detailed technical migration instructions and notes, etc., are entered in the Software and/or Hardware Implementation Steps sections.

PRODUCTION TURNOVER APPROVAL FORM

A development team cannot simply install the product in the production environment and hand it over to the operations team to support. There must be a transition period where the development team helps ensure the product is properly installed, configured, and working fully in conformance with the requirements and technical specifications. The transition period and services are specified in the warranty agreements.

Before the warranty period expires, it's important that the customer, and more likely their operations support team, signs off that they have approved the delivery of the software and agree that the product and its implementation conform to the requirements and the technical specifications. When the deployment is to an internal customer, the approval process allows the development team to move on to new work. When the deployment is to external customer, the agreement limits legal liabilities and also provides evidence to allow invoicing the customer for all remaining nonpaid services and deliverables.

The Production Turnover Approval process includes a number of reviews, such as:

- Verification that the systems and code were tested and conform to requirements and specifications
- Verification of proper installation and configuration within the target production environment
- Verification that all systems and code were reviewed and approved by key development and operations personnel
- Verification that systems migration activities were successful
- Verification that any problems or bugs found during production were resolved or mitigated
- Verification that backup and recovery procedures work as documented
- Other verifications, as necessary

The Production Turnover Approval template is a simple form that contains the following sections:

- Introduction
    - Purpose
    - Project / System Overview
    - Scope
- Production Turnover Approval Request
- Approval and Signoff

YOUR COMPANY LOGO

Production Turnover Approval-PTA
Project Name
Version

- *Internal or external interfaces.*

2 Production Turnover Approval Request

Provide project and/or system PTA request information in the following table.

Topic	Description
Request Date	
PTA Requester's Name and Contact Information	
Project or System Number	*Include associated numbers, if any*
Project and/or System Name	
Application and/or Module Name, if any	
Device Service Type and Name, if any	
PTA Type	*e.g., Major, minor, significant, emergency and whether a new or deleted system, upgrade, fix, patch, relocation, configuration modification*
Approval Priority	*i.e., Urgent, high, low, normal*
Project Manager's Name and Contact Information	
Requested Production Turnover Date and Time	
Project / System / Application Description	
Turnover Risk	*i.e., High, medium, low*
Associated Files, if any	*i.e., File name and date*
Fix Change Information, if applicable	*Provide root cause analysis and solution*
Other Relevant Information	

 Document: 8050 Page 5 of 6

www.SDLCforms.com

This page is intentionally left blank

Project Closure / Maintenance Phase
Overview

As projects come to an end, there is still work that needs to be done. This is true for both agile and traditional-based software development projects. In theory, agile-based projects can go on for an infinite amount of time. However, in actual practice, the needs of customers and users begin to diminish, or business needs change, and corporate IT funds are allocated to other higher priority programs and projects. Also, the external contractors assisting with the development of their client's products can be replaced at any time, regardless of whether they are supporting an agile-based or traditional software development project.

When this happens, the projects need to be formally closed out in such a manner that the client and development organizations properly maintain the physical or intellectual property assets developed by or devoted to the project, plus the team needs to ensure the retention of all lessons learned. In addition, the employees assigned to the effort need to be reassigned, while contractors need to be offboarded and have their access rights removed.

If the project related work is for an external customer, the customer needs to formally accept the deliverables of the project. This is important to ensure the contract development organization no longer has any financial or legal liabilities, and to ensure the delivery organization gets paid for their effort.

The closeout phase may run in parallel with a warranty period. So even as the project team is working toward leaving the project, they may still be on the hook to fix bugs and address minor change requests.

As a quick summary, the final close phase of the software or systems development project has the following goals and objectives:

- Effectively transition the solution to the receiving organization
- Ensure the organization has a record of all critical design, development, production support, infrastructure, and security data on all components of the solution
- Ensure transfer of product maintenance and support efforts to the receiving operations personnel
- Ensure end users are properly trained and using the product effectively
- Obtain formal acceptance of the system from the customer
- Capture both project and product-oriented lessons learned during the project
- If requested, implement a formal change management process to capture new customer's desired enhancements for future development activities
- Implement an effective Disaster Recovery Program
- Implement a process to retire the solution when the system no longer meets organizational needs
- Before personnel leave and start new projects, have them complete a Knowledge Transfer Document to capture many facets of the design for subsequent maintenance.

The methods and tools documented in this chapter support the above goals and objectives.

Forms & Templates

LESSONS LEARNED TEMPLATE

Regardless of whether a project is judged a success or not, there is a lot to learn from each project. In fact, there may be more to learn from the activities that didn't work out so well than there is from the activities that exceeded expectations. At the completion of every activity, it's useful for the participants to ask three questions:

1. What well, and why?
2. What didn't work so well, and why?
3. How can we do things better in the future?

In the traditional software development model, the team captures lessons learned as part of the close process. However, if the participants take the time to document their answers to these questions at the completion of each activity, then it will take less time and effort to record the lessons learned at the end of the project.

The practice of analyzing work and documenting methods for improvement is so powerful that the techniques are a critical requirement in most agile-based software development practices. An example is the Sprint Retrospective Events within the Scrum Framework. The questions asked in the retrospectives can be the same three noted above, but there are also other questioning formats in use, such as:

- **What should we start, stop or continue doing?**
- **What made us mad, sad or glad?**
- **"4 Ls" - What did we like, learn, lack and long for?**
- **What was good, what was bad, any new ideas, any required actions?**

The lessons learned template shown on the next page is a program or project level summary document that implements the following sections and subtopics:

- **Project close-out discussions**
  - Attendees
  - Project's biggest successes
- **Lessons learned that worked well or didn't work well:**
- **Knowledge areas**
  - Project initiation
  - Project planning
  - Project management
  - Staffing
  - Project communications
  - Project funding
  - Project costs
  - Schedule compliance
  - Project roles and responsibilities
  - Risk management
- **Knowledge areas**
  - Procurement
  - Requirements definition
  - Scope creep
  - Development
  - Quality assurance testing
  - Training
  - Documentation
  - Approvals

YOUR COMPANY LOGO

Lessons Learned Template
Project Name
Version

Project Close-Out Discussions	
Attendees	
Project's Biggest Successes	

List areas of potential improvement as well as high-impact improvement strategies.

Lessons Learned				
Knowledge Area		**Lessons Learned That Worked Well**		**Lessons Learned That Didn't Work Well**
Project Initiation	1.		1.	
	2.		2.	
	3.		3.	
Project Planning	1.		1.	
	2.		2.	
	3.		3.	
Project Management	1.		1.	
	2.		2.	
	3.		3.	

TRANSITION OUT PLAN

The Transition Out Plan, or "Project Management Transition Plan," details the activities, communications, schedules, roles and responsibilities, and approvals processes necessary to move a product and its related artifacts from control of the development team over to the receiving operations team. The transition plan further describes how the project deliverables will be brought to full operational status and integrated into ongoing operations and subsequently maintained.

The purpose of the transition activities is to ensure that all deliverables provide the business value that justified the development efforts. In other words, rather than a simple delivery and standup of the solution components, the dev teams go further to ensure the operations teams can effectively support the new project, and that the product is fully in compliance with the requirements that justified the corporate investments.

The Transition Out Plan shown on the following page contains the following sections and subtopics:

- Transition Approach
- Transition Plan Objectives
- Transition Team Organization
- Transition Process Tasks
- Knowledge Transfer Process
- Product Delivery
  - Rollout Plan
  - Data Migration
- Communication Plan
- Transition Schedule
- Handover and Acceptance.

Transition Out Plan
Project Name
Version

Note: Text displayed in blue italics is included to provide guidance to the author and should be deleted before publishing the document. In any table, select and delete any blue line text; then click Home→Styles and select "Table Text" to restore the cells to the default value.

1 SUMMARY

The Transition Out Plan is used to describe how project deliverables will be brought to full operational status, and integrated into ongoing operations and subsequently maintained. Its purpose is to ensure that these deliverables will be used effectively to produce the requisite Business Value after project completion.

The Transition is a sum of the work to be done to create an effective support apparatus.

Include in this section a succinct summary including a top-level description of the transition plan at a high level and what the plan should accomplish. This section should include an overview and history of the agreement, specifically who the agreement pertains to, who it is transitioning to, and the timeframe/period of transition and any facts which are needed to successfully complete the transition.

2 TRANSITION APPROACH

This section discusses the overall approach and length of the transition. It includes the identification of any functions that still need to be performed and items which must be considered such as staff realignment during the transition or additional staff to handle and manage the transition. State any assumptions that you have made that may have made critical decisions in the document.

3 TRANSITION PLAN OBJECTIVES

Briefly describe the objectives of the plan.

For example:

- *Ensure that deliverables are of adequate quality to support full Business Value realization.*
- *Ensure that the customer is fully trained, capable of effectively using the deliverables, and willing to use the deliverables.*
- *Ensure that the work environment will support effective use of the deliverables.*
- *Identify staffing and training needs for system operation and maintenance.*
- *Facilitate ongoing updates to deliverables, ongoing help desk operations, and Quality Assurance oversight.*

POST PROJECT REVIEW SURVEY QUESTIONNAIRE

In larger companies, projects report into a Program Management Office (PMO). The PMO is a resource to help implement skilled resources and approved project management processes across the organization. During the Project Closure / Maintenance phase of a project, the PMO should conduct a survey with executives, team members and stakeholders to gather feedback on the completed project.

The survey helps ensure proper transitioning of the project's products and provides a reusable guide to gather important information that can help improve performances on future projects. General questions are asked of the stakeholders and to determine how well the team performed against the constraints of the project. (e.g., scope, schedule, budget, resources and quality.)

This survey helps the PMO gather and assess the thoughts and perspectives of those individuals most affected by the project, but external to the project team. Recipients of the survey questions can answer each question by selecting from a range of responses from "Strongly Agree" to "Strongly Disagree."

The Post Project Review Survey Questionnaire, shown on the following page, includes the following sections, each containing a number of pre-defined questions, to organize and document responses to the post project review questions:

- General project issues
- Project communications
- Scheduling and estimating
  - Project phase schedules
  - Actual completion dates
- Design and implementation
- Test processes
- Training and documentation
- General process questions
- Project costs
  - Project cost items budgeted vs. actual costs
- Approval section.

YOUR COMPANY LOGO

Post Project Review
Project Name
Version

During the Project Closure / Maintenance phase of a project, the Project Management Office (PMO) conducts a survey to gather feedback on the project to improve performance on subsequent projects. This survey will assist the PMO in gathering project sponsors and team member's thoughts and perspectives on the project.

Your replies to this survey will be kept strictly confidential.

Note: In any table, select and delete any blue line text; then click Home→Styles and select "Table Text" to restore the cells to the default value.

Post Project Review

Section 1: General Project Issues

Question	Strongly Agree	Agree	Neither Agree or Disagree	Disagree	Strongly Disagree	Not Applic -able
How clearly defined were the objectives for the project?	☐	☐	☐	☐	☐	☐
How clearly defined were your objectives for your work tasks?	☐	☐	☐	☐	☐	☐
How clear were you as to your role in the project?	☐	☐	☐	☐	☐	☐
How adequately involved did you feel you were in the project?	☐	☐	☐	☐	☐	☐
How well do you believe that executive management supported this project?	☐	☐	☐	☐	☐	☐
Do you feel appreciated, recognized and rewarded for your efforts?	☐	☐	☐	☐	☐	☐
	☐	☐	☐	☐	☐	☐
	☐	☐	☐	☐	☐	☐

INSTALLATION BACK-OUT ROLLBACK PLAN TEMPLATE

Best plans sometimes go awry, and a newly installed system may crash or otherwise fail, not perform to specifications, or improperly implement key business and user capabilities. In those situations, the new system must be rolled back while gracefully bringing the original applications back online with a snapshot of data that is most current and not corrupted.

Installation Back-Out Rollback Plan Template documents instructions for the following:

- **System requirements including download and extract procedures, database information, installation and CRON scripts, system configuration and steps necessary to fine tune the database.**
- **Implementation steps including steps to set up user accounts.**
- **Back-out procedures including the back-out strategy and considerations.**
- **A description of the rollback process with the inclusion of considerations, criteria and risks.**

An Installation Back-Out Rollback Plan Template, such as the one pictured on the following page, may contain the following sections and subtopics:

- **Purpose**
- **System Requirements**
  - Platform Installation and Preparation
  - Download and Extract Procedure
  - Database Creation
  - Installation Scripts
  - Cron Scripts
  - System Configuration
  - Database Tuning
- **Implementation**
- **Back-out Procedure**
  - Back-out Strategy
  - Back-out Considerations
    - Load Testing
    - User Acceptance Testing
  - Back-out Criteria
    - Back-out Risks
  - Back-our Authority
  - Back-out Procedure
- **Rollback Procedure**
  - Rollback Considerations
  - Rollback Criteria
  - Rollback Risks
  - Rollback Authority
  - Rollback Procedure

Installation, Back-Out and Rollback Plan
Project Name
Version

2.2 Download and Extract Procedure

Provide the chronological steps for downloading and installing the software.

#	Step	User
1		
2		
3		
4		
5		

2.3 Database Creation

Provide the instructions for creating the database.

#	Database Step	User
1		
2		
3		
4		
5		

2.4 Installation Scripts

Identify all of the installation scripts in the order in which they are executed.

#	Installation Script	User
1		
2		
3		
4		
5		

PROJECT ACCEPTANCE DOCUMENT

A project is not technically closed until the external or interior customer has signed off that they are satisfied that all project related products, artifacts and other requirements were delivered to specifications. The specifications include the deliverables specified in the original contracts or project charters - that approved and initiated the project, plus all requirements and specifications delivered under the terms of the originating contract or approval agreements.

This is the final step in the Close process after an application goes "Live," and is important to ensure that your customers are satisfied with the project outcomes, to remove legal liabilities, and to provide proof to an external client's Payments Department that final funds can be released.

The Project Acceptance Document includes a listing and brief descriptions of the products and services the customer received. The document along with approval signatures formalizes the acceptance of the project and its deliverables.

The Project Acceptance Document, as shown in the next page, contains the following sections and subtopics:

- **Topic**
  - Project Name
  - Project Number
  - Department or Business Unit Name
  - Department or Business Unit Cost Center
  - Sponsor Name and Telephone Number
  - Project Manager Name and Telephone Number
- **Project Description**
- **Statement of Acceptance**
- **Signatures**

YOUR COMPANY LOGO

Project Acceptance Document (PAD)
Project Name
Version

Note: Text displayed in blue italics is included to provide guidance to the author and should be deleted before publishing the document. In any table, select and delete any blue line text; then click Home→Styles and select "Table Text" to restore the cells to the default value.

This form formalizes acceptance of the <Name of Project> Project.

Topic	Description
Project Name	
Project Number	
Department or Business Unit Name	
Department or Business Unit Cost Center	
Sponsor Name and Telephone Number	
Project Manager Name and Telephone Number	

Project Description
Description of the products and services produced by the project:

Statement of Acceptance:

The products and services produced by this project satisfactorily meet all specifications and requirements that have been documented for the project. The products and services are completely developed, have been adequately tested, and are ready for use in ongoing operations.

Note: Business area signoff on the Project Acceptance Document indicates acceptance of this responsibility for the software delivered by the project.

Signatures:

Sr. Business Mgr:		Date:	
Sponsor/Business Owner:		Date:	
Project Manager:		Date:	
Process Analyst:		Date:	

www.SDLCforms.com

MODIFICATION / CHANGE CONTROL REQUEST FORM

Once a product is released into a production environment, and put to use, it's not unusual to have the customers and users come up with new requirements. Very often, having access to the system in an operational mode provides new insights on how the product can be used and improved. In the traditional software development model, a new project must be approved and funded to support such requests.

The Modification/ Change Control Request Form helps develop and maintain the list of desired enhancements along with information to help justified the continued development activities. Each request requires an evaluation to determine the proposed scope of work, technical feasibility, capabilities, the features and functions to implement the capabilities, and time, cost and resource estimates. The feasibility assessments help limit potential impact to business capabilities and/ or IT operational capacity, architecture, infrastructure, compliance, and schedules. But they also provide the cost/ benefit information executive managers need to justify new investments in the existing software applications or systems.

Change requests are typically requested for the following reasons:

- To fix identified bugs and defects
- To document newly identified requirements
- Identify opportunities to integrate systems to reduce redundant data entry across multiple systems
- To update the applications to run on new computing equipment, operating systems, networks, or other infrastructure changes
- To improve availability, reliability and scalability
- To implement more robust security systems and controls

The Modification/ Change Control Request Form, as shown on the next page, contains the following sections and subtopics:

- Reason for the Change
- Description
- Assumptions
- Project Impacts
  - Estimated Schedule Impacts
  - Breakdown of Estimated Hours and/or Costs
  - Capital / Expense Worksheet
- Approvals
-

YOUR COMPANY LOGO

Modification/Change Control Request (MR) #
Project Name
Version

Complete the following table as a guide to explain the estimated additional capitalized and expensed costs associated with this modification request.

CAPITAL / EXPENSE WORKSHEET			
Project Work Order #			
	Estimate	Estimate	
Internal IT Resources	Staff Cost	Capital	Expense
Subtotal	$0	$0	$0
External Professional Services	Professional Services Cost	Capital	Expense
Subtotal	$0	$0	$0
Hardware / Communications & Maintenance	Cost	Capital	Expense
Subtotal	$0	$0	$0
Software Licenses & Maintenance	Cost	Capital	Expense
Subtotal	$0	$0	$0
Other (Travel, Supplies, etc)	Cost	Capital	Expense
Subtotal	$0	$0	$0
Software Licenses	Cost	Capital	Expense
Subtotal	$0	$0	$0
	Cost	Capital	Expense
SUBTOTALS	$0	$0	$0
Contingency	Cost	Capital	Expense
Contingency (15%)	$0	$0	$0

 Document: 9100 Page 5 of 6

www.SDLCforms.com

DISASTER RECOVERY PLAN INFORMATION TEMPLATE

Natural and human-induced disasters happen. They must be planned for to reduce the severity of the impacts when they occur. Disaster recovery plans include policies, procedures and tools to ensure data is properly protected, and have backup systems and plans to recover use of the applications as quickly as possible.

Two of the most important parameters are mean time between failure (MTBF) and mean time to repair (MTTR). MTBF includes the system design, infrastructure, and facilities considerations plus daily actions taken to keep the system from failing in the first place, and to make sure all data is properly backed up at all times. MTTR includes the strategies to quickly identify the root cause and locations of the system failures, resolve the issues, and bring the systems back online. Larger data centers, often with thousands of servers, make the process of problem identification, location identification, and resolution much more challenging.

The Disaster Recovery Plan template is used to build a system's disaster recovery plan. In effect, a disaster recovery plan is a contingency plan to address issues related to system failures. A disaster recovery plan is required for all publicly traded companies but are also employed by government agencies and other commercial companies that need to minimize their losses or liabilities that occur when the organization is unable to support its business function with the loss of their critical information systems.

The Disaster Recovery Plan template, as shown in the following page, contains the following sections:

- What is disaster recovery?
- Goals, objectives, and scope
- Disaster recovery team
- Disaster recovery time
- Disaster recovery site
- Critical services or information
- Technology priority for services and applications
- Response process (notice of problem, assessment, and resolution)
- Disaster recovery declaration
- Notification and recovery procedures
- Network, Internet and Windows Server recovery plan
- Electronic mail and telecommunications recovery plan.

YOUR COMPANY LOGO

Disaster Recovery Plan Information
Project Name
Version

2.2 Disaster Recovery Time

Recovery time to start and maintain operations is vital to the financial needs of the company. The following are sample categories and tiers to begin operations, services, and technology in the event of a disaster: The organization needs to define those items.

Disaster Recovery Category	Recovery Time
Infrastructure	*0-8 Hours*
Mission Critical or Tier One	*8-24 Hours*
Mission Essential or Tier Two	*24 - 48 Hours*
Business Unit Essential or Tier Three	*48 - 72 Hours*
Business Unit Essential or Tier Four	*120+ Hours*
New applications not yet assigned a criticality ranking.	*To be determined; defaults to priority 4 until management approval is received of a different priority*

2.3 Disaster Recovery Site

A disaster recovery site should be pre-configured specifically for disaster circumstances. The location of the disaster recovery site is: (Enter location here).

2.4 Critical Services or Information

Provide a methodology to ensure that mission critical services are identified. See the following table with example information.

Service or Information	Description
Revenue Dependent Technology Services	*Services where the company's revenue would be directly impacted if certain technology services are not available are given the highest priority. Data required for these services will be copied to the disaster recovery site in real time in the event of an emergency.*
Productivity Services	*These technology services allow the company to work more productively and save operating expenses. These services will be prioritized from an economical standpoint.*
Technology Infrastructure	*These are the basic technology services required by business operations such as redundant electrical, cooling, network, security, internet access, telephones, email, etc.*
Employee Offices	*Offices space for "X" number of employees at the disaster recovery site. The site will be configured for easy remote access from home*

 Document: 9150 Page 7 of 25

CERTIFICATE OF COMPLIANCE AND ACCEPTANCE OF DELIVERABLE FORM

The Certificate of Compliance and Acceptance of Deliverable form is another though simpler version of the project acceptance document. This form is used to close out deliverable items and is therefore used across the duration of a project. The review and approvals verify the customer has reviewed and accepted the specified deliverable items. The use of these forms throughout the project makes the final project close activity simpler, as the project team has the documentation showing progressive successful deliveries. Contractors use this form to document customer acceptance of each deliverable so that they can get paid for the effort.

The Certificate of Compliance and Acceptance of Deliverable form is generally used to validate customer acceptance of project deliverables in accordance with requirements specified in the project's charter, or customer contract agreements, or task orders or purchase orders.

The Certificate of Compliance and Acceptance of Deliverable form, as shown in the following page, includes the following sections and subtopics:

- **Contractor / Independent section to be completed**
  - Task order number
  - Deliverable number
  - Deliverable name
- **Project manager section to be completed**
  - Task order number
  - Deliverable number
  - Deliverable name
- **Contracting office section to be completed**
  - Task order number
  - Deliverable number
  - Deliverable name
  - Contractor's project manager signs and dates the form.

YOUR COMPANY LOGO

Certificate of Compliance and Acceptance of Deliverable
Project Name
Version

This Certificate of Compliance is generally used to accept and validate project deliverables provided by outside contractors and developers in accordance with a task order or purchase order, but can be used in any situation where you wish to review the status of deliverables by internal organizations on a given project.

Note: In any table, select and delete any blue line text; then click Home→Styles and select "Table Text" to restore the cells to the default value.

Certificate of Compliance and Acceptance of Deliverable

Contractor / Independent Consultant Completes This Section			
Task Order No.		Deliverable No.	
Deliverable Name			
I hereby certify that the deliverable complies with all contractual requirements and specifications.			
Contractor Project Manager	*Signature*	Date	*Date*

Project Manager Completes This Section			
Task Order No.		Deliverable No.	
Deliverable Name			
The above deliverable has been received and inspected. The deliverable meets the contractual requirements, specifications and minimum quality standards stated in the task order. I recommend that this deliverable be accepted in its entirety.			
Deliverable Passed inspection	*Date*	Inspected By	
Contractor Project Manager	*Signature*	Date	*Date*

Contracting Officer Completes This Section			
Task Order No.		Deliverable No.	
Deliverable Name			
The deliverable stated above, contracted for in the above referenced project, is hereby accepted. Contractor may invoice for this deliverable.			
Contractor Project Manager	*Signature*	Date	*Date*

 Document: 9220 Page 1 of 1

www.SDLCforms.com

REQUEST FOR NEW APPLICATION / ENHANCEMENT FORM

Organizational departments and other groups may, from time-to-time, require construction of new applications, extensions to existing applications, or procurement of commercial-off-the-shelf (COTS) applications. The Request for New Application / Enhancement form facilitates the requirements gathering and approval process associated with new product requests.

The overarching organization, be it a commercial, government or nonprofit entity, needs a process to document these requirements and review them in terms of the organizations approved Technical Reference Model (TRM) and Standards Information Base (SIB). The organization's TRM provides a model and taxonomy listing generic platform services. The SIB is a database of identified corporate IT standards. Approvals are granted based on needs, budgets, and conformance with the organization's approved standards.

The Request for New Application / Enhancement form, as shown on the following page, implements the following sections and subtopics to provide the information that is necessary to initiate a new request.

- Requestor / Business sponsor
- Applicable department
- Contact person
- Request Type
  - New application
  - Enhancement to existing application
  - Minor correction
- Description of request
- Priority (Critical, high priority, low priority, delete)
- Potential risks
- Funding sources qualified
- Related projects
- Attachments

The Request requires the following entries to be completed by the project manager / development manager:

- Estimated development hours
- Anticipated delivery date
- Conclusions
- Approvals.

YOUR COMPANY LOGO

Request for New Application/Enhancement to Existing Application
Project Name
Version

Request for New Application / Enhancement to Existing Application							
Requestor / Business Sponsor							
Applicable Department							
Contact Person							
Request Type	☐	New Application	☐	Enhancement to Existing Application	☐	Minor Correction	
Application Name				Need Date			
Description of Request							
Priority	☐ Critical	☐ High Priority		☐ Low Priority		☐ Delete	
Potential Risks							
Funding Source Qualified	☐ Yes	☐ No					
Related Projects							
Attachments							
To be Completed by Project Manager / Development Manager							
Estimated Development Hours				Anticipated Delivery Date			
Conclusions:	☐	To be discussed at next status meeting					
	☐	Project delayed – Resources not available					
	☐						
Approvals							
Project Manager					Date		
Development Manager					Date		

PRODUCT RETIREMENT PLAN TEMPLATE

Eventually, all software applications and systems either lose their value, become obsolete, or become unsupportable or unsustainable. When the time come to retire a system (a.k.a. end of life), care must be taken to ensure the organization does not cause problems for the organization. For example,

- Communicate intentions to all affected business units and users well in advance of the retirement and disposal
- Develop and execute plan to migrate data and functionality to newer/ replacement systems
- Ensure all necessary business processes and operational functionality previously implemented by the legacy system are supported in the new environment
- Ensure data that must be retained is archived or moved into another business system, as appropriate
- Ensure any data no longer required for retention is properly disposed of, included shredding the original data storage devices, tapes and disks
- Ensure all legacy equipment is properly disposed of to prevent loss of data and to protect against environmental problems

The Product Retirement Plan provides a detailed roadmap to retire a software product or system. The plan includes descriptions of the related hardware, software and data storage components associated with the system or application, how they will be removed from production, what data must be archived or migrated to another system, plus the planned retention or disposal policies governing legacy information. The product retirement plan also documents how and when users and support personnel must be notified prior to retiring the system.

The Product Retirement Plan as shown on the following page contains the following sections:

- Product or system information
- Reason for retirement
- Costs and benefits
- Any assumptions
- Dependencies or constraints
- List of stakeholders
- Risks
- Implementation dates.

YOUR COMPANY LOGO

Product Retirement Plan
Project Name
Version

Note: Text displayed in blue italics is included to provide guidance to the author and should be deleted before publishing the document. In any table, select and delete any blue line text; then click Home→Styles and select "Table Text" to restore the cells to the default value.

1 Purpose

Purpose describes the intent of the document, which is to provide detailed instructions for retirement of the (product or system name). It includes the following information:

- *Retirement Strategy: How the hardware, software, data, and documentation associated with the product or system will be detached from production and archived or migrated.*
- *Communication: Information that will be sent to users and support personnel to notify them to retire the system and associated activities.*

2 Product or System Information

This section of the document should contain background information about the product or system to be retired.

2.1 Project or System Overview

Provide the purpose of the system or product and the environment in which it is currently used.

Indicate the products, systems, and processes that interface with it.

Provide diagrams, if necessary.

2.2 Reason for Retirement

Provide information why the product or system is being retired, e.g., old technology, new compliance laws, it's old, too costly, new process).

2.3 Cost and Benefits

Provide information about the costs associated with the retirement and the benefits that will result from the retirement.

 Document: 9300 Page 4 of 10

OFF-BOARDING CHECKLIST

Regardless of whether a project is managed under traditional or agile-based software development practices, when the work is no longer required, the employees and contractors working on the development effort must be reassigned or let go. The offboarding process must facilitate activities that are in compliance with organizational, program/ project, human resource, and legal requirements. Some common areas that must be addressed by the offboarding process include:

- Communications
- Employee reassignments or terminations
- Employee termination benefits
- Letters of termination
- Contract terminations
- Knowledge transfer
- Transition work activities and period
- Nondisclosure and noncompete agreements (if applicable)
- Exit interviews
- Recovery of organizational assets
- Remove network, system, email, and facility access rights, as applicable
- Other

This Offboarding checklist template provides a guide on the activities that must be completed during the offboarding process. The checklist should be usable for multiple situations, such as contract and project endings, retirement, death or disabilities, employee transfers, and employee or contractor terminations.

The Off-Boarding Checklist shown on the next page includes the following sections and subtopics:

- Exit information:
  - Employee name, position and number
  - Work phone, email address, department
  - Manager/supervisor, work phone and email address
  - Last day of work and type of attrition.
- If promoting/transfer:
  - Future department, position and manager/supervisor and phone number.
  - Deliverable number
  - Deliverable name
- Employee actions
- Supervisor / manager actions
- Department personnel staff actions
- Information technology actions
- Equipment checklist
- Employee name, signature and date.

YOUR COMPANY LOGO

Off-Boarding Checklist
Project Name
Version

Exit Information							
Employee Name:							
Position Name:					**Position Number:**		
Work Phone:				**Email Address:**			
Department:							
Manager/ Supervisor							
Work Phone:				**Email Address:**			
Last Day of Work: (in current position)	**Mo.**	**Day**	**Year**	**Type of Attrition:**	☐	**Retirement:**	**(Date)** 99/99/99
	99	99	99		☐	**Transfer**	
					☐	**Promotion**	
					☐	**Termination**	

If promoting/ transferring:			
Future Department:		**Future Position:**	
Future Manager/ Supervisor:		**Manager/Supervisor Phone Number:**	

	Employee Actions	Date Completed
☐	Notify Manager/Supervisor of intent to leave/change employment w/ at least two week's notice.	
☐	Submit a letter of resignation (if separating employment) to Manager/Supervisor.	
☐	Return all company-issued equipment on or before the last day of work/position.	
☐	Complete the Exit Survey/Questionnaire, prior to his/her last day of service.	
☐	Participate in an exit interview with Human Resources Manager, or other party.	
☐	If applicable, contact Retirement.	
☐	If applicable, contact Benefits.	

GLOBAL APPLICATION SUPPORT SUMMARY TEMPLATE

The organization's IT department may find it's useful to summarize and maintain a record of important product information, in a single document, for every product within their portfolio. For example, the document might maintain a record of critical design, development, production support, infrastructure and security related information. The document serves as a relatively brief reference guide, as opposed to going out through a multitude of project and product documents to find a specific detail or better understanding of how the product was designed, developed, supported and deployed. This is the purpose of the Global Application Support Summary Template.

The Global Application Support Summary, a sample of which is shown on the next page, spans 30 pages and contains the following sections:

Section 1 – Applications Data

Section 2 – Design and Development/Integration

Section 3 – Production Support

Section 4 – Infrastructure

Section 5 – Security

Section 6 – Instructions for Completing Index.

YOUR COMPANY LOGO

Global Application Support Summary
Application Name
Version

2. Application Data

Application Name				
Description				
Business Use				
Application Type	☐ Application	☐ Interface	☐ Commercial-Off-The-Shelf (COTS)	
Vendor Name (If applicable)				
CONTACTS				
Contact Type	Name	Phone #	Cell #	E-Mail Address
Primary Technical Contact				
Secondary Technical Contact				
Business Owner				
Business Analyst				
Production Support Contact				
BUSINESS PROFILE				
Business Group				
Cost Center				
Application	☐ Tier 1 ☐ Tier 2 ☐ Tier 3 ☐ Tier 4			
Application Active?	☐ Yes ☐ No	SOX Application?	☐ Yes ☐ No	

NOTES:

DEVELOPER KNOWLEDGE TRANSFER REPORT

Regardless of whether they are contactors or employees, developers come and go over the lifecycle of a product. When a product needs to be enhanced there is no guarantee that the original developers are still available to work on the enhancements, or even still employed by the organization. On traditional software development projects, the concern is heightened as there may be breaks between new enhancement development activities. Therefore, it's critical that steps are taken to have the developer record information.

The Developer Knowledge Transfer Report implements a method and a tool to document important technical and implementation decisions made during the development of the product.

The primary purpose of the Developer Knowledge Transfer Report is to transfer knowledge from the initial development staff to new developers who make take over modifications, bug fixes and enhancement in future development projects. The Developer Knowledge Transfer Report can be used by the sustainment team within the operations department.

The Developer Knowledge Transfer Report, as shown on the next page, includes the following sections and subtopics:

- **References**
- **Key Personnel**
  - Business users
  - Subject matter experts (SMEs)
  - Developers
- **Technical knowledge**
  - Core languages
  - Dependent languages
  - Reporting tools
  - Databases
  - Operating systems
- **Business Knowledge**
- **Application knowledge**
  - Application level function
- **Application environment**
  - Flows / diagrams
  - Server-side application components
  - Client-side application components
  - User – network environment.

YOUR COMPANY LOGO

Developer Knowledge Transfer Report
Project Name
Version

The Developer Knowledge Transfer Report provides a vehicle for conveying details about a system or application for production support developers. This document will provide the support for developers by transferring knowledge for the application from the initial development people to new developers with precise data about the project development.

Note: Text displayed in blue italics is included to provide guidance to the author and should be deleted before publishing the document. In any table, select and delete any blue line text; then click Home→Styles and select "Table Text" to restore the cells to the default value.

1 Purpose

The purpose of this document is to ensure that the developer is prepared and equipped with the knowledge necessary to successfully support and maintain the application in an efficient and timely manner.

- **Section 2, Key Personnel**, identifies the business users, subject matter experts and developers who contributed to the application.
- **Section 3, Technical Knowledge**, identifies the programming, database, and reporting skills necessary for a developer to be successful in supporting and maintaining the application.
- **Section 4, Business Knowledge**, identifies the client business processes that relate to the application.
- **Section 5, Application Knowledge**, details technical design information specific the application
- **Section 6, Application Environment**, details the network environment for the application.

1.1 Scope

This Knowledge Transfer Report is a living document and will be updated whenever functionality is added, changed, enhanced or removed from the application.

1.2 Terminology

The table below defines terms and acronyms that are necessary in understanding the application.

Term	Definition

Confidential – ©2015 Documentation Consultants (www.SDLCforms.com) Document: 2200 Page 4 of 14

www.SDLCforms.com

Index

A

Acceptance Criteria 34, 36, 44, 58, 130, 138, 158, 174, 208
Action Item (Status) 234
Activity Worksheet 138
Adapt Phase 61
Adaptive Software Development 70
Agile 15-23
  Agility 19, 20
  Agile Manifesto 15, 17, 18, 19, 57, 70-73
Agile Project Management 21, 58
  APM 21, 33
Agile Tools and Templates 17-68
Analysis and Design Document 190
as-is assessment 98, 158, 159

B

BABOK Guide 159
Backlog Status 46-49
Back-Out Rollback Plan 274
Bug List 224
Bug Report 226
Building our Digital World 13, 14, 16, 20
  BODW 14
burndown chart 64
burnup chart 62
Business case 23, 76, 78, 80, 84, 86, 88, 102, 136
Business Process Modeling Notation 184
  BPMN 184
Business Requirements Document 15, 21, 88, 164, 166
  BRD 88, 164, 166

C

Capability Maturity Model Integration 21
  CMMI 21, 138, 298
Capacity Sheet 50, 52
capital 120, 166, 278
Change Control 254, 278
Change Management 110, 112, 231, 232, 260, 267
Close Phase 68, 267
COBIT 144, 146
Code Review Checklist 204
commercial-off-the-shelf 140, 284
  COTS 140, 285
Concept of Operations 74, 76, 88, 89
  CONOPS 74, 76, 88, 90
concept phase 20, 21, 23, 76, 77, 91, 102, 148, 158
Configuration Management 122
CM 122
Contact Log 132
Conversion Plan 206
Cost Estimating 116, 142
Cost-Benefit Analysis 76, 86, 148
Crystal 70
current state 158

D

Data Flow Diagrams 184
  DFD 184
Database Design Document 196, 197
Deployment Phase 249-251
Developer Knowledge Transfer Report 292, 293
Development Estimating Worksheet 118, 119
Development Operations 19, 20,68
  DevOps 19, 20,68,
Disaster Recovery Plan 154, 182, 186, 188, 267, 280, 281
Dynamic Systems Development Method 70
  DSDM 70

E

Envision Phase 23, 28
Epics 15, 21, 33, 34-37, 40, 54
expense 120, 121, 166, 278
Explore Phase 67
extreme programming 70
  XP 70

F

Feasibility Study 76, 82, 83, 102, 136
Features 19, 26, 33, 35, 36, 40, 42, 44, 58, 72, 84, 88, 148, 158, 159, 170, 184, 188, 195, 214, 220, 222, 256, 278
Five Ws and How 252, 260
Framework 21, 33, 42, 108, 142, 144, 184, 268
Functional Modeling 184, 185
functional requirements 158, 168, 169, 170, 174, 184, 186, 188, 194, 209, 220
Functional Requirements Document 15, 21, 136, 158, 168, 169
future state 26, 158

I

initiate project checklist 102, 103
Installation Planning Guide 250, 254, 255
Integrated Definition (IDEF) 184

integrated development environments 16
IDE 16
International Institute of Business Analysis 298, 159
IIBA 298, 159
Issue 15, 24, 28, 30, 33, 38, 42, 54, 57, 70, 76, 80, 82, 86, 94, 96, 107, 110, 130, 134, 139, 154, 158, 162, 164, 184, 186, 194, 216, 231, 238, 240, 244, 246, 250, 272, 280
iterative and incremental development 19
IID 19, 20

J
Jim Highsmith 18, 21

L
Lessons Learned 56, 61, 68, 130, 231, 267-269
List of Opportunities 156, 157

M
Maintenance Phase 75, 267, 272
Manager's Weekly Checklists 154, 155
mean time between failure 280
MTBF 280
methods and tools 14-18, 21, 23, 38, 61, 68, 69, 70, 71, 73-76, 107, 110, 158, 174, 184, 209, 267

N
Non-functional requirements 158, 168, 174, 184, 186, 194, 209, 220, 228

O
Off-Boarding 288, 289
On-Boarding 100, 101
open source 16, 186,
Operations Guide 136, 250, 260, 261

P
Planning Poker 38-40
process guide 136, 250, 252, 253
Process Modeling 184, 185
Procurement Plan 128, 129
Product Backlog Grooming 54, 55
product backlog refinement 33,42, 54, 55
product owner 16, 21, 23, 26, 30, 33, 42, 44, 54, 57, 58, 76, 77, 78, 88, 91, 92, 94, 98, 132, 140, 154
Product Retirement Plan 286, 287
Production Implementation Plan 250, 262, 263
PIP 250, 262, 264
Production Turnover 136, 250, 264, 265
Products 14, 15, 18, 19, 20, 21, 42, 58, 70-72, 74, 78, 82, 84, 96, 110, 122, 158, 208, 209, 216, 260, 267, 272, 276
Project Acceptance 276, 277, 282
Project Administration Phase 231, 233
Project Charter 24, 25, 72, 76, 78, 79, 80, 91, 102, 136, 154, 160, 276
Project Closure 154, 267, 272
Project Cost Estimate 116
Project Data Sheet 30, 31
Project Initiation 75, 76, 91, 92, 93, 102, 103, 110, 268
Project Management Office 108, 109
PMO 108, 109, 132, 272
Project Management Plan 110, 110
Project Management Transition Plan 270, 271
Project Milestone Status Form 30, 78, 231, 242, 243
Project Organization Chart 132, 133
Project Plan Schedule Template 152, 153
Project Planning 33, 75, 91, 102, 107, 122, 130, 185, 268
Project Resource Plan 91, 104, 105
Project Status 56, 96, 231, 234, 244, 245
project team 28, 30, 33, 68, 76, 78, 91, 92, 96, 97, 98, 102, 104, 107, 108, 114, 122, 136, 138, 154, 185, 226, 234, 236, 238, 267, 272, 282
Projects 4, 15, 16, 18, 19, 21, 24, 61, 66, 68, 70-78, 86, 87, 98, 108, 110, 130, 132, 134, 138, 140, 160, 185, 204, 209, 231, 232, 236, 246, 267, 272, 284, 292

Q
QA Checklist 210, 211
Quality Assurance 16, 107, 110, 130, 132, 140, 172, 208, 209-211, 216-218, 242, 268
Quality Control 130, 130
Quality Plan 130, 131

R
RACI 134, 136
RASCI 134
Regression Testing 208, 228, 229
Release Plan 33, 42, 42
Report Design 202, 203
Request for Information 148, 149
RFI 148, 150
request for quote 112, 148
RFQ 112, 148
Requirements Change Impact Analysis 176, 177
Requirements Inspection Checklist 172, 173
Requirements Phase 158, 159
Requirements Traceability Matrix 136, 174, 175
RTM 28, 174, 175
Retrospective 56, 57, 68
Risk 24, 28, 29, 30, 33, 70, 71, 75-77, 80, 82, 86, 91, 94, 96, 104,107, 110, 112, 116, 122, 124-128, 134, 144, 150, 154, 160, 164, 168, 170, 176, 192, 200, 202, 206, 220, 222, 228, 231, 236-238, 240, 244, 246, 254, 268, 274, 284, 286
Risk Analysis 33, 86, 124, 126, 127

Risk Management Plan 110
Risk Management Log 28, 29
Risk Management Register 235, 236
Roles and Responsibilities Matrix 134, 135, 162
Root Cause Analysis 124, 150, 151

S
Sarbanes-Oxley Act 144, 146
  SOX 144, 147
Scrum 50, 56-58, 68, 70, 268
Scrum Master 16, 42, 44, 57, 58
Service Level Agreement 182, 183
  SLA 182, 183
Software Architecture Plan 186, 187
  SAP 186, 187
Software User Guide 136, 250, 256, 257
Speculate Phase 33, 57
Sprint Backlog 33, 34, 36, 40, 46-53
Sprint Planning 40, 44-57
Stakeholder Identification 91, 98, 99
Standards Information Base 284, 285
  SIB 284, 286
start phase 91, 102
Statement of Work 112, 113
  SOW 112, 114
Stories 15, 21, 33-46, 54-50, 62, 174
Story Points 39-45, 54-56, 62, 64, 66
System Administration Guide 250, 258, 259
System Design Document 136
System Design Phase 184, 185
system quality assurance 216, 217
  SQA 216, 218
system test plan 220, 221
Systems Requirements Specifications 188, 189
SRS 188, 190

T
Taxonomy 74, 75, 284
Technical Reference Model 284
  TRM 284
Technical Requirements Document 194, 195
  TRD 194, 196
Test Cases 212, 222
Test Plan 136, 209, 212, 214, 215, 220, 221, 228
Test Scenarios 18, 172, 208, 209, 212, 213, 222
Test Scripts 72, 208, 212
Testing Phase 192, 208, 209
Themes 21, 34-37, 40, 54
to-be state 88, 158, 159
Training Plan 136, 178, 179
Training Request 180, 181
Transition O

ut Plan 270, 271

U
UAT 214, 222, 223
Unified Modeling Language 170, 171, 185,
  UML 170, 185
Use Case 15, 170, 171, 186, 194
User Acceptance Test 136, 214, 222, 223
  UAT 214, 222, 224
User Interface (UI) 118, 184, 190, 196, 198, 200, 201, 209, 256

V
value proposition 76, 80, 84, 85, 86
Velocity Chart 61, 66, 67
Vision Statement 24, 26, 27, 31

W
WBS decompositions 142
WBS Resource Planning template 140, 141
Website Planning Checklist 198, 199
Website Testing Summary Template 209, 218, 219
Work Breakdown Structure 138-143
work in process 20, 158
WIP 20

Bibliography

1 Beck, K., Beedle, Mi. Bennekum, A., Cockburn, A., Cunningham, W., Fowler, M. Grenning, J., Highsmith, J. Hunt, A., Jeffries, R., Kern, J., Marick, B., Martin, R., Mellor, S., Schwaber, K., Sutherland, J., Thomas, D. (2001) The Agile Manifesto. http://agilemanifesto.org/ Accessed 30 January 2020

2 Beck, K., Beedle, Mi. Bennekum, A., Cockburn, A., Cunningham, W., Fowler, M. Grenning, J., Highsmith, J. Hunt, A., Jeffries, R., Kern, J., Marick, B., Martin, R., Mellor, S., Schwaber, K., Sutherland, J., Thomas, D. (2001) Independent Signatories of The Manifesto for Agile Software Development. http://agilemanifesto.org/display/index.html Accessed 30 January 2020

3 Beck, K., Beedle, Mi. Bennekum, A., Cockburn, A., Cunningham, W., Fowler, M. Grenning, J., Highsmith, J. Hunt, A., Jeffries, R., Kern, J., Marick, B., Martin, R., Mellor, S., Schwaber, K., Sutherland, J., Thomas, D. (2001) History: The Agile Manifesto. http://agilemanifesto.org/history.html Accessed 30 January 2020

4 Beck, K., Beedle, M., Van Bennekum, A., Cockburn, A., Cunningham, W., Fowler, M., Grenning, J., Highsmith, J. Hunt, A., Jeffries, R., Kern, J., Marick, B., Martin, R.C., Mellor, Steve., Schwabver, K., Sutherland, J., Thomas, D. (2001). Manifesto for Agile Software Development. Retrieved from www.agilemanifesto.org/principles.html Accessed 30 January 2020

5 Highsmith, J. (2010) Agile Project Management, Second Edition. Pearson Education, Inc. Boston, MA

6 Corporate (2015). A Guide to the Business Analysis Body of Knowledge® (BABOK® Guide). Version 3. 2015. Toronto, Ontario, Canada: International Institute of Business Analysis (IIBA).

Acknowledgments

This book would not have been possible without the collaboration of Robert 'Bob' DeLaet, Managing Director of Documentation Consultants (a.k.a. Chief Cook and Bottlewasher) and co-founder of SDLCforms. (www.sdlcforms.com/) Through the hours of work over many months, and the stress of writing a book, Bob's humor got us through the project, and I will value his friendship for life.

As a side note, anyone who has worked in a small company environment understands everyone must pitch in and perform multiple tasks; hence Bob's subtitle of 'Chief Cook and Bottlewasher.' Now, most people would not have a problem with someone showing humor of that nature, and so Bob added the subtitle to his business cards when he moved his business from California back to his home in the eastern part of the nation. However, a manager of one of his clients took offense to the title, thinking it was too unprofessional for someone they wanted to work with, and so they pulled his contract. While Bob views the situation as a learning experience, he didn't allow it to affect his sometimes-quirky sense of humor, which made working with him so much more interesting and fun.

I also want to thank Joe McDaniel of BookCrafters (http://www.bookcrafters.net/) for stepping in to help with proofreading the final drafts, and Mary Walewski of Buy The Books Marketing (https://buythebookmarketing.com/) for helping me develop our marketing strategies and programs for this series of books. James Neely and Victory Moore of Unzipped Designs (https://www.unzippeddesign.com/) helped me keep my sanity by taking the burden of developing my website and the artwork for the Building Our Digital World series covers off my plate. I've come to trust their technical knowledge and artistic instincts implicitly. And, finally, I'm forever thankful to Adam Hay of Adam Hay Studio (https://www.adamhaystudio.com/) for taking on the design work for the books in this series. I'm confident he didn't know what he was getting into, given the expanded scope of this project and the numerous iterations to get things right. His guidance, steady hand, and patience allowed this project to keep moving forward.

Made in United States
North Haven, CT
05 November 2026

21764043R00167